# Advances in Metal Processing

# SAGAMORE ARMY MATERIALS
# RESEARCH CONFERENCE PROCEEDINGS

**Recent volumes in the series:**

20th: **Characterization of Materials in Research: Ceramics and Polymers**
Edited by John J. Burke and Volker Weiss

21st: **Advances in Deformation Processing**
Edited by John J. Burke and Volker Weiss

22nd: **Application of Fracture Mechanics to Design**
Edited by John J. Burke and Volker Weiss

23rd: **Nondestructive Evaluation of Materials**
Edited by John J. Burke and Volker Weiss

24th: **Risk and Failure Analysis for Improved Performance and Reliability**
Edited by John J. Burke and Volker Weiss

25th: **Advances in Metal Processing**
Edited by John J. Burke, Robert Mehrabian, and Volker Weiss

# Advances in Metal Processing

**Edited by**

## John J. Burke
*Army Materials and Mechanics Research Center*
*Watertown, Massachusetts*

## Robert Mehrabian
*National Bureau of Standards*
*Washington, D.C.*

and

## Volker Weiss
*Syracuse University*
*Syracuse, New York*

PLENUM PRESS · NEW YORK AND LONDON

Library of Congress Cataloging in Publication Data

Sagamore Army Materials Research Conference, 25th, Bolton Landing, N.Y., 1978.
  Advances in metal processing.

  (Sagamore Army Materials Research Conference proceedings; v. 25)
  Includes bibliographical references and index.
  1. Metal-work—Congresses. 2. Metallurgy—Congresses. I. Burke, John J. II. Mehrabian, Robert. III. Weiss, Volker, 1930-    . IV. Title. V. Series: Sagamore Army Material Research Conference. Proceedings; v. 25.
  UF526.3.S3 vol. 25 [TS200]           623'.028s [671]              81-439
  ISBN 0-306-40651-9

Proceedings of the Twenty-Fifth Sagamore Army Materials Research Conference,
held July 17–21, 1978, Sagamore Hotel, at Bolton Landing, Lake George, New York.

©1981 Plenum Press, New York
A Division of Plenum Publishing Corporation
233 Spring Street, New York, N.Y. 10013

All rights reserved

No part of this book may be reproduced, stored in a retrieval system, or transmitted,
in any form or by any means, electronic, mechanical, photocopying, microfilming,
recording, or otherwise, without written permission from the Publisher

Printed in the United States of America

# SAGAMORE CONFERENCE COMMITTEE

Chairman
**DR. J. J. BURKE**
Army Materials and Mechanics Research Center

Secretary
**MR. JOSEPH A. BERNIER**
Army Materials and Mechanics Research Center

Conference Coordinator
**HELEN BROWN DEMASCIO**

Program Committee

**DR. VOLKER WEISS**
Syracuse University

**DR. J. J. BURKE**
Army Materials and Mechanics Research Center

**PROF. MERTON C. FLEMINGS**
Massachusetts Institute of Technology

**DR. ROBERT MEHRABIAN**
National Bureau of Standards

**MR. FRANCIS C. QUIGLEY**
Army Materials and Mechanics Research Center

**DR. GEORGE MAYER**
Army Research Office

**PROF. NICHOLAS GRANT**
Massachusetts Institute of Technology

PREFACE

Syracuse University and the Army Materials and Mechanics Research Center of Watertown, Massachusetts have conducted the Sagamore Army Materials Research Conference since 1954. In celebration of the 25th Anniversary of this conference, these proceedings are dedicated to the founding members of the Sagamore Conferences. They are Prof. Dr. George Sachs, Dr. James L. Martin, Colonel Benjamin S. Mesik, Dr. Reinier Beeuwkes, Mr. Norman L. Reed and Dr. J. D. Lubahn.

This volume, ADVANCES IN METAL PROCESSING, addresses Rapid Solidification Processing, Powder Processing and Consolidation, Welding and Joining, Thermal and Mechanical Processing, Metal Removal and Process Modeling.

The dedicated assistance of Mr. Joseph M. Bernier of the Army Materials and Mechanics Research Center and Helen Brown DeMascio of Syracuse University throughout the stages of the conference planning and finally the publication of this book is deeply appreciated.

Syracuse University
Syracuse, New York                                              The Editors

# CONTENTS

## OVERVIEW

Materials Processing – A Perspective of the Field . . . .  1
   M.C. Flemings and R. Mehrabian

### SESSION I

#### RAPID SOLIDIFICATION PROCESSING
#### B.B. Rath, Moderator

Heat Flow Limitations in Rapid Solidification
     Processing . . . . . . . . . . . . . . . . . . . .  13
   R. Mehrabian, S.C. Hsu, C.G. Levi, and
   S. Kou

Laser Processing of Materials . . . . . . . . . . . . . .  45
   B.H. Kear, E.M. Breinan, and E.R. Thompson

Electrohydrodynamic Techniques in Metals
     Processing . . . . . . . . . . . . . . . . . . . .  79
   J. Perel, J.F. Mahoney, B.E. Kalensher,
   and R. Mehrabian

### SESSION II

#### POWDER PROCESSING AND CONSOLIDATION
#### A.M. Adair, Moderator

Fundamentals of Particulate Metallurgy . . . . . . . . . .  91
   A. Lawley

### SESSION III

#### WELDING AND JOINING
#### F.C. Quigley, Moderator

Welding with High Power Lasers . . . . . . . . . . . .  111
   E.M. Breinan and C.M. Banas

## SESSION IV

### THERMAL AND MECHANICAL PROCESSING
G. Mayer, Moderator

Fundamentals of Superplasticity and Its
    Application . . . . . . . . . . . . . . . . . . . . . .   133
  O.D. Sherby, R.D. Caligiuri, E.S. Kayali,
    and R.A. White

Advances in the Heat Treatment of Steels . . . . . . . . .   173
  J.W. Morris, Jr., J.I. Kim, and C.K. Syn

## SESSION V

### METAL REMOVAL
J.J. Burke, Moderator

Innovations in Grinding Materials . . . . . . . . . . . .    215
  R.A. Rowse and J.E. Patchett

Recent Advances in Grinding . . . . . . . . . . . . . . .    229
  P. Guenther Werner

Mathematical and Economic Models for Material
    Removal Processes . . . . . . . . . . . . . . . . . .    257
  Vijay A. Tipnis

## SESSION VI

### PROCESS MODELING
L. Croan, Moderator

Modelling Macrosegregation in Electroslag
    Remelted Ingots . . . . . . . . . . . . . . . . . . .    277
  D.R. Poirier, M.C. Flemings, R. Mehrabian,
    and H.J. Klein

The Analysis of Magnetohydrodynamics and Plasma
    Dynamics in Metals Processing Operations . . . . . .     319
  C.W. Chang, J. Szekely, and T.W. Eagar

Computer Simulation of Solidification . . . . . . . . . .    345
  William C. Erickson

Index . . . . . . . . . . . . . . . . . . . . . . . . . .    377

# MATERIALS PROCESSING - A PERSPECTIVE OF THE FIELD

M. C. Flemings* and R. Mehrabian
Massachusetts Institute of Technology
M. C. Flemings* and R. Mehrabian

Massachusetts Institute of Technology
Cambridge, MA

University of Illinois
Urbana, IL

## INTRODUCTION

The materials activities of the past few decades have helped to crystallize two concepts that are important in considering the future. One of them is the total materials cycle; the other is the field of materials science and engineering. A third, yet to be fully crystallized is materials processing. The latter can be viewed as the transition steps linking the five stages of the materials cycle, as well as, that part of materials science and engineering that links basic research to the solution of practical problems.

In this overview presentation metals processing is treated within this broader definition of the field of materials processing. Its' role within this broader spectrum is commensurate with the important role of metals in our economy and should be so recognized.

## PROCESSING AND THE MATERIALS CYCLE

The future for materials will depend in large measure on our response to forces which are now well recognized. Some of these are the classical laws of supply and demand; others include such national problems as energy, environment, regulation, productivity, and such international problems as the rapidly changing third world nations. Together these forces constitute great opportunities

*Is now with the National Bureau of Standards, Washington, DC.

for the field of materials science and engineering, and particularly for materials processing.

The materials cycle is a physical concept - materials flow from the earth through various useful forms and back to the earth in a closed system that is global in extent. Materials science and engineering (MSE) is an intellectual concept - a coherent system of scientific and engineering disciplines that combines the search for insights into matter with the use of the resulting knowledge to satisfy society's needs for materials. MSE in its broadcast dimensions overlays the materials cycle and can be employed to adjust it to solve specific materials problems. Materials processing is the development and synthesis of concepts of materials science and engineering, and their combination with the art and technology of the fields to produce in new or improved processes, and in new or improved properties and products through new processes.

The total materials cycle is shown in highly simplified form in Figure 1. The cycle is driven by societal demand, and materials traverse it in five stages; these five transitions comprise the field of Materials Processing (MP):

- Extraction of raw materials: ores and minerals, rock, sand, timber, crude rubber.
- Processing of raw materials into bulk materials: metals, chemicals, cement, lumber, fibers, pulp, rubber, electronic crystals.
- Processing of bulk materials into engineering materials: alloys, ceramics and glass, dielectrics and semiconductors, plastics and elastomers, concrete, building board, paper, composites.
- Fabrication of engineering materials into structures, machines, devices, and other products.
- Recycling discarded products (materials) to the system or returning them permanently to the earth.

The materials cycle provides an analytical framework for dealing with upsets in the availability of materials at various points in the cycle, whether with reference to the world, a nation, an industry, a company, or a factory. The cycle is a real system of interacting subsystems. The flow of materials at a given point, therefore, can be sensitive to economic, political, and social decisions made at other points. Analyzed in this context, materials shortages usually are found to be due not to scarcity in the world, but to dislocations in the cycle that interfere with the arrival of materials at a given point in the usual amounts and at reasonable

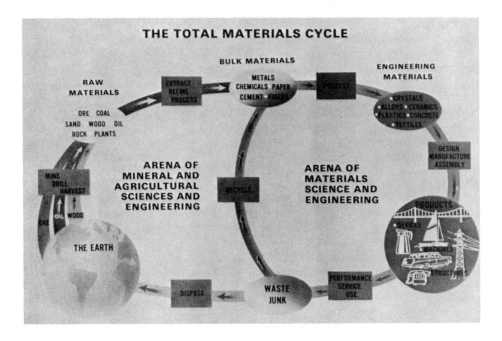

Figure 1. The Total Materials Cycle (1).

prices. A shortage may arise at one point, for example, because of inadequate processing capacity at another point.

Materials, energy, and the environment interact strongly at virtually every point in the materials cycle. The energy required to produce refined copper, for example, rises sharply as the copper content of the ore declines. At the same time, more and more rock and gangue must be disposed of per unit of copper produced, which throws further burdens on the landscape. Indeed, almost all materials processing affects the environment in one way or another, adding social costs to the other costs of moving materials through the cycle.

About one third of the energy consumed by industry in the United States goes into the value added to materials by production and fabrication. On the other hand, materials are crucial to making energy available in the first place. In fact, inadequacies in the performance of materials currently are the primary constraint on the efficiency, reliability and safety, cost-effectiveness, or even actual realization of essentially all of our advanced energy-conversion technologies - gas turbines, nuclear reactors, high-energy-density batteries, fuel cells, magnetohydrodynamics, coal conversion, and solar-energy conversion.

The total materials cycle on the whole can be considered an essential analytical tool in the development of national materials policies in both industrialized and developing nations. Although the cycle is in concept independent of the rates of flow of materials, by its very nature it is intertwined with the lines of supply and demand. The flow of materials around the loop can be disturbed drastically and unpredictably by events outside the domain of materials. Nevertheless, steps can be taken within the cycle to prepare appropriate countermeasures, including stockpiling, recycling, and substitution of one material for another.

The flow of materials from one point in the cycle to the next (i.e., the processing of materials) depends, as indicated above, on technical, economic, political and social factors. Thus, the intellectual arena of Materials Processing, while residing firmly in technology, must include as well economics, political science, and social aspects.

## PROCESSING AND MATERIALS SCIENCE AND ENGINEERING

Earlier, materials science and engineering was described as a coherent system of scientific and engineering disciplines. The central purpose of MSE (Figure 2) is twofold: to probe the relationships among the internal structure of materials and their properties and performance; to use the resulting knowledge in producing, shaping, and otherwise processing materials so as to control their properties and achieve the desired performance in the finished product. The second part of this twofold purpose, Materials Engineering is primarily but not exclusively Materials Processing. Materials Processing provides the links between science art and technology in producing, shaping and otherwise treating materials. Materials Engineering in addition includes such tasks as materials specification, failure analysis, etc.

At its most ambitious level, MSE links fundamental understanding of the behavior of electrons, atoms, and molecules to the performance

# MATERIALS PROCESSING - A PERSPECTIVE OF THE FIELD

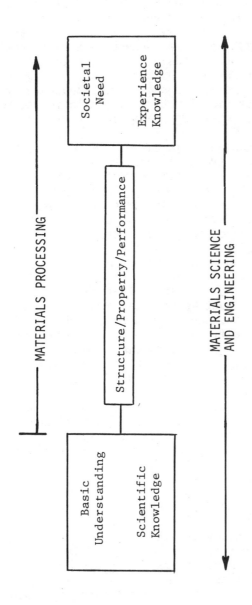

Figure 2. Materials Processing as Part of Materials Science and Engineering.

of products. Materials Processing as a part of MSE links basic research to the solution of practical problems.

MSE relies on a two-way flow of information between the empirical practitioner at one extreme and the basic scientist at the other. For millenia man has been accumulating empirical knowledge of materials. But only in this century - and at an accelerating pace in the past few decades - have scientists begun to acquire the corresponding basic insights. The microstructure of materials has been revealed by the microscope, crystal structure by x-ray diffraction, molecular and atomic structure by various spectroscopies, electronic structure by excitation techniques, nuclear structure by high-energy radiations.

With this new knowledge, it has become possible by design to exploit the linkage of structure, properties, and performance. The strength and dimensional stability of polymers, for example, can be upgraded through methods of synthesis that yield highly ordered molecules that cluster into crystalline order. Transistors are made by manipulating the electronic structure of a particular class of solids (semiconductors); they are produced on a large scale through methods of processing that achieve exceptionally precise control of composition and internal structure.

Studies of MSE at work indicate that the two-way flow of information is most productive when basic understanding of a materials problem and the empirical need to solve it are mixed so intimately that it is difficult to tell which provided the initial impetus toward a solution. In the main, however, the initial impetus in the MSE system seems to arise more often from "societal pull" than from "scientific push". The successes of the MSE approach to materials should not be construed to mean that properties can necessarily be predicted from structure alone, nor performance from properties, except in a relatively general sense. As a rule, the structure-properties-performance relationships must be worked out through the reciprocal flow of information - scientific and empirical - across linkages that comprise the intellectual body we call Materials Processing.

## MATERIALS PROCESSING

Much of the technology of materials that relates to civilization's prosperity, security, and quality of life is encompassed by the field of Materials Processing. Materials Processing includes bulk processing operations such as materials beneficiation, chemical processing, refining, recycling, and processes for changing shape and controlling properties, such as forging, casting, glass blowing, plastic injection molding, and vapor deposition. It also includes broad systems problems associated with materials production and utilization - those at the level of a plant or industry, and those which can be dealt with only at a national or global level. It comprises specifically the five stages of transition in the materials cycle, Figure 1, which are shown more directly in Figure 3.

- ο Raw material extraction
- ο Processing raw materials
- ο Processing bulk materials
- ο Fabrication of engineering materials
- ο Recycling

An alternate way to view Materials Processing is as in Figure 2, comprising a unifying, applications oriented, linkage to the field of MSE, in the development and synthesis of concepts of MSE (and their combination into the art and technology of other fields) to produce new or improved processes, and new or improved properties and products through new processes.

The chart of Figure 4 provides a more detailed perspective of the components of Materials Processing. The field deals with technical, economic, and social aspects of connecting a raw material to a semi-finished material or finished component. "Structure - Property - Performance" are central themes, as are "Costs and Benefits", social and economic.

The Body of Knowledge Comprising Materials Processing

One way to divide the field of Materials Processing is into five "phase related" groupings as follows:

Table 1: Classes of Processing

Vapor

Liquid

Solid

Vapor-liquid

Vapor-solid

Liquid-solid

Vapor-solid processes, for example comprise an important part of refining of many refractory metals, and of vapor deposition. Liquid-solid processes are important in casting, crystal growing and liquid phase sintering.

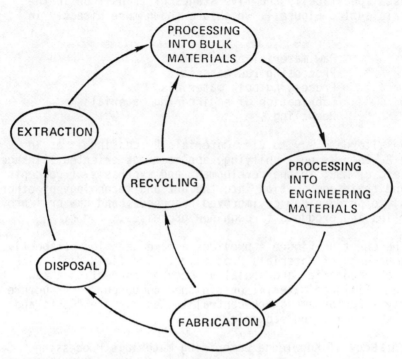

Fig. 3. Materials Processing As Transition Steps in the Materials Cycle.

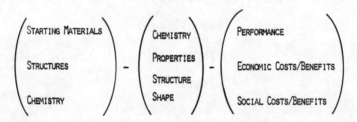

Fig. 4. Materials Processing.

An alternate, and equally useful way of conceptualizing the field of Materials Processing is by the change that is sought:

<u>Table 2: Classes of Processing</u>

<u>Composition Change</u>

Alloying
Refining

<u>Structure Change</u>

Phase change
Polymerization

<u>Shape Change</u>

Rheological processes
Chip forming processes
Consolidation processes

Here, processes are divided as to whether the change being created is one of composition, structure, or shape.

Either of the categorizations above would promote the basis for university courses in Materials Processing. Both, however, fall far short of adequately delineating the body of knowledge that comprises Materials Processing, or of delineating the essential components of a successful materials processing program.

Figure 5 is an attempt to show schematically this broader base of Materials Processing. In describing Figure 5, it will be helpful to refer to a specific process innovation, and the one we will choose is one that is not yet a practical reality, but is much talked about - strip casting of steel.

A large market exists for economic thin gauge steel sheet (for example for automobile manufacture) and current technology is relatively costly and energy consuming compared with an ideal continuous process that would produce strip close to the final desired thickness so only moderate reduction (e.g. 3/1) would be necessary in rolling. Some fundamental materials processing problems stand in the way of developing such a process, including the problem of strip cracking, and achieving adequate surface finish. Substitution of these problems will require building on a strong Engineering Science base, and synthesizing this base with art and the specific engineering technologies involved. At each step in the development of this process, the economic and social (e.g. environmental) tradeoffs must be appropriately considered.

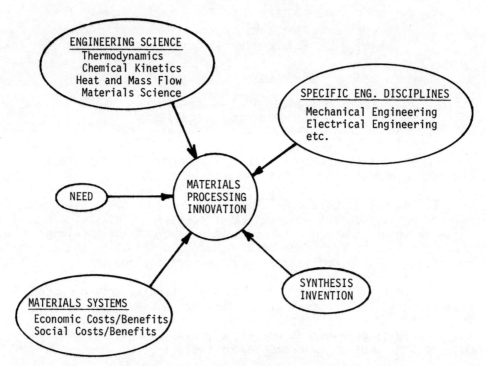

Fig. 5. The Broad Base of Materials Processing Innovation.

In short, the development of a process innovation such as the example above, requires an interdisciplinarity of effort that would include a broad segment of the materials community and also extends well beyond it to include faculty from other disciplines, industry and government personnel. A very long list of similar such programs could easily be developed. Following are some examples:

Table 3:  Examples of Potential Process Developments

    Metal matrix composites

    Electronic gauging

    Ultra-homogeneous structures

    Supersaturated structures

    Nonequilibrium structures

    Fine particle materials

    Automation/robotization

    Ultra low energy processing

    Single crystal turbine blades

    Continuous single crystal growth

    Strip casting of steel

    Light weight materials for automobiles

    New methods of processing in space

## REFERENCES

1. Materials and Man's Needs, Vol. 1, "The History, Scope and Nature of Materials Science and Engineering," National Academy of Sciences, Washington, DC, 1975.

# HEAT FLOW LIMITATIONS IN RAPID SOLIDIFICATION PROCESSING

R. Mehrabian, S.C. Hsu, C.G. Levi, and S. Kou

Department of Metallurgy and Mining Engineering
Department of Mechanical and Industrial Engineering
University of Illinois
Urbana-Champaign, Illinois

## INTRODUCTION

The term Rapid Solidification Processing, RSP, is equally applicable to the formation of both crystalline and non-crystalline solid phases by quenching of a material from an initial liquid state. During RSP cooling rate in the liquid prior to solidification affects nucleation (undercooling) and growth phenomena in important ways — it influences undercooling in crystalline solidification and is an overriding factor in the formation of non-crystalline structures. On the other hand, the fineness of a crystalline microstructure (e.g. segregate spacing, size of second phase particles, etc.) can usually be correlated to average cooling rate during solidification or time available for coarsening. Thus, a clear distinction must be made between cooling rates in the liquid (or during non-crystalline solidification) and during crystalline solidification; the latter is significantly lower at equivalent rates of external heat extraction due to the heat of fusion.

In what follows some general relationships are presented between cooling rates during crystalline and non-crystalline solidification and process variables in different RSP techniques. Calculations are presented to show the heat flow <u>characteristics and limitations in the three general areas of RSP; atomization and solidification against substrates with and without significant resistance to heat flow at the liquid-substrate interface.</u>

## HEAT FLOW DURING ATOMIZATION

During solidification of small spherical alloy droplets, heat flow is controlled by both convection at the surface and by radiation. However, there are no accurately established values for the combined radiative and convective heat transfer coefficient, and direct measurement of the cooling rate or heat flux during solidification of an atomized droplet would be extremely difficult, if not impossible. In gas atomization the convective heat transfer coefficient is overriding and is usually estimated from the following equation:

$$\frac{hD}{k_f} = 2.0 + 0.60 Re^{1/2} Pr^{1/3} \qquad (1)$$

where:

$Re$ = Reynold's number = $vD\rho_f/\mu_f$

$Pr$ = Prandtl number = $C_{pf}\mu_f/k_f$

$C_{pf}$ = specific heat of the gas

$D$ = particle diameter

$k_f$ = conductivity of the gas

$h$ = heat transfer coefficient

$v$ = gas velocity relative to particle

$\rho_f$ = density of the gas

$\mu_f$ = viscosity of the gas

An upper limit on achievable heat transfer coefficents can be deduced from equation (1). For example, the calculated heat transfer coefficients during argon gas atomization, with a high relative velocity of one Mach between the gas and the metal droplets, are $5.86 \times 10^3$ and $1.1 \times 10^4$ W/m$^2$.K for droplet diameters of 75μm and 25μm, respectively. Using higher conductivity gases and finer particles result in calculated heat transfer coefficients of less than $10^5$ W/m$^2$.K.

Indirect estimates of heat transfer coefficients in various atomization processes have also been made by comparison of measured segregate (dendrite arm) spacings in crystalline alloy powders with predetermined relationships between these spacings and average cooling rates during solidification. Table I shows the various heat transfer coefficients during atomization of Maraging 300 steel determined by this method. Note that the heat transfer coefficient for gas atomization is the same order of magnitude as that estimated above from equation (1).

In general, then, a limitation on the achievable heat transfer coefficient at a liquid metal droplet — environment interface can be translated into a limitation on the important dimensionless variable - Biot Number - governing the rate of heat extraction from the droplet. For example, a heat transfer coefficient $h < 10^5$ W/m$^2$·K translates to a limitation on the range of Biot Numbers of $10^{-2} <$ Bi $< 1.0$ for atomized droplets of liquid aluminum in the size range of 1μm to 1000μm.

$$Bi = \frac{hr_0}{k_\ell} \qquad (2)$$

where h is the heat transfer coefficient at the metal droplet-environment interface, $r_0$ is the radius of the droplet and $k_\ell$ is the conductivity of the liquid metal.

For smaller heat transfer coefficients, Biot numbers below 0.01, Newtonian cooling expressions are generally considered to be applicable. However, as shown below, Biot numbers should be below 0.001 before temperature gradients in a liquid droplet become negligible. For Bi > 0.001 numerical heat flow models are necessary. A short summary of achievable cooling rates and solidification times, from one such numerical heat flow model, based on the limitations noted above is presented below from a recent paper on solidification of atomized droplets (1).

Figure 1 shows calculated dimensionless temperature distribution in a liquid droplet for various Biot Numbers and an initial superheat of $T - T_g/T_M - T_g = 1.3$ at the instant the droplet surface reaches its melting point. These data show that for Biot Numbers less than ∼0.001 there is no significant temperature gradient in the droplet and the simple Newtonian cooling expressions can be used for crystalline and non-crystalline solidification. On the other hand, the results also indicate that even for small Biot Numbers, in the range of ∼0.01, there may be significant temperature gradients in a metal droplet. For example, in a 20μm diameter droplet of liquid iron where Bi ∼0.01 (h - $4 \times 10^4$ W/m$^2$ K) the maximum temperature gradient, at the droplet surface when the surface reaches the melting point, is ∼ $1.5 \times 10^6$ K/m and a temperature difference of 7.5K between the surface and the center of the droplet is calculated. Thus, the Newtonian cooling assumption that temperature

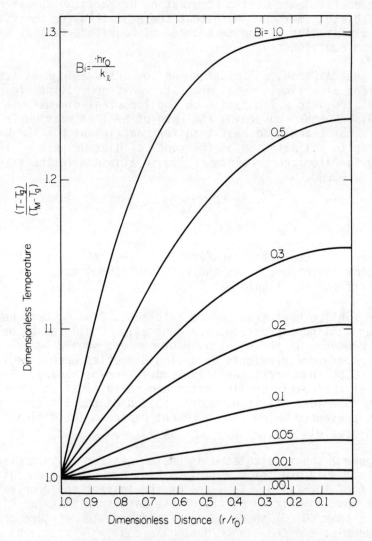

Figure 1. Dimensionless temperature distributions in a liquid droplet when its surface reaches the melting temperature $T_M$, for different Biot numbers and dimensionless initial superheat $T-T_g/T_M-T_g = 1.3$ (from Reference 1).

TABLE I

CALCULATION OF HEAT TRANSFER COEFFICIENTS FROM DAS
(From Reference 2)

MARAGING 300 STEEL

$$d = 39.8 \, \varepsilon_{Avg}^{-0.30}$$

| ATOMIZATION PROCESS | PARTICLE SIZE, μm | DAS μm | $\varepsilon_{Avg.}$ °K/sec | h(CALCULATED) c.g.s. | h(CALCULATED) S.I. | $\frac{hr_o}{k_\ell}$ |
|---|---|---|---|---|---|---|
| ARGON ATOMIZED FINE POWDER | 75 | ~2 | ~2.1x10$^4$ | ~0.23 | 9.6x10$^3$ | 0.0084 |
| REP | 170 | ~3 | ~5.5x10$^3$ | ~0.13 | 5.4x10$^3$ | 0.011 |
| STEAM ATOMIZED COARSE POWDER | 1000 | ~6.5 | ~4.2x10$^2$ | ~0.06 | 2.5x10$^3$ | 0.029 |
| VACUUM ATOMIZED | 650 | ~6.5 | ~4.2x10$^2$ | ~0.039 | 1.63x10$^3$ | 0.0123 |

d = secondary dendrite arm spacing, DAS.

$\varepsilon_{Avg}$ = average cooling rate during solidification.

Biot number = $\frac{hr_o}{k_\ell}$

differences inside a body are negligible for Bi ⩽ 0.01 may not always be justified in atomization processes.

An important variable effecting undercooling prior to crystalline solidification or formation of amorphous structures is the cooling rate in the liquid droplet. A generalized expression relating the dimesionless instantaneous average cooling rate, $\varepsilon_{avg}$, in a liquid metal droplet to the Biot Number and dimensionless surface temperature has been derived (1):

$$\varepsilon_{avg} = 3 \times Bi \; (\theta_{SURFACE} + Ste)$$

where:

$$\varepsilon_{avg} = \frac{r_o^2 \, C_{ps}}{\alpha_\ell \, \Delta H_{s\ell}} \left(\frac{\partial T}{\partial t}\right)_{avg} = -3 \int_0^1 \phi^2 \left(\frac{d\theta}{dFo}\right) d\phi \quad (3)$$

$C_{ps}$ = specific heat of the solid

$Fo = \dfrac{\alpha_\ell t}{r_o^2}$ = Fourier number or dimensionless time

$\Delta H_{s\ell}$ = Heat of fusion

$Ste = C_{ps} (T_M - T_g)/\Delta H_{s\ell}$ = Stefan number

$t$ = time

$T$ = temperature of the droplet

$T_g$ = temperature of the environment

$T_M$ = melting point of the droplet

$\alpha_\ell$ = thermal diffusivity of the liquid metal

$\theta = \dfrac{C_{ps}(T - T_M)}{\Delta H_{s\ell}}$ = dimensionless temperature

$\phi = r/r_o$ = fractional radius

Note that if thermal properties of a non-crystalline solid are assumed to be equal to the liquid the above expression would apply throughout solidification and subsequent cooling of amorphous droplets.

Expression (3) indicates that average cooling rate, $(\partial T/\partial t)_{avg}$, in a liquid metal droplet is directly proportional to the heat transfer coefficient at the droplet-environment interface and inversely proportional to the radius of the droplet. Thus, considering an upper limit for achievable heat transfer coefficients the only other method for increasing cooling rate is to decrease particle size. For example, using $h = 10^5$ W/m$^2$·K and expression (3) the average cooling rates of $8.3 \times 10^5$ K/s and $8.3 \times 10^6$ K/s are calculated for iron droplets of 100μm and 10μm, respectively, when their surface reaches the melting point.

Figure 2 shows the effect of process variables on the normalized crystalline solidification time of three different metal droplets. This time is normalized by dividing it by the Newtonian prediction of net solidification time which is:

$$\left. \frac{\alpha_\ell t_f}{r_o^2} \right|_{Newtonian} = [3 \, Bi \, Ste \, (C_{p\ell}/C_{ps})]^{-1} \quad (4)$$

Crystalline solidification time is the most important variable affecting segregate spacings, inclusion size, etc. It gives a better indication of time available for coarsening phenomena to occur than average cooling rate during solidification. The general trends established in Figure 2 for this normalized solidification time indicate that once heat transfer at the droplet-environment interface is maximized this time can be reduced by decreasing particle size. Furthermore, increasing initial superheat prolongs solidification, and the effect is larger as the Biot Number increases; the sensible heat retained in the liquid portion of the droplet increases resulting in longer times from initiation to completion of solidification. For example, 500μm droplets of iron solidifying with an $h = 5 \times 10^4$ W/m$^2$K (Bi = 0.31) will require $4.8 \times 10^{-3}$ seconds to complete solidification with no initial superheat and $5.4 \times 10^{-3}$ seconds with a dimensionless initial superheat of 0.1 (∼150K). Finally the effect of superheat on net solidification times diminishes with decreasing Biot Numbers. For example, for the same heat transfer coefficient, 50μm droplets of iron (Bi = 0.031) would solidify in $2.6 \times 10^{-4}$ seconds and $2.7 \times 10^{-4}$ seconds for initial superheats of zero and 150K, respectively.

Figure 3 shows dimensionless solid-liquid interface velocity as a function of Biot Number for various positions of the interface in iron droplets solidifying with concentric isotherms. The data show that dimensionless interface velocity increases with Biot Number and the progress of solidification.

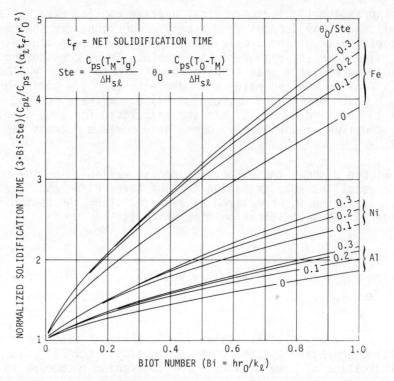

Figure 2. Normalized net solidification time for liquid droplets of aluminum, iron and nickel, as a function of Biot number and dimensionless initial superheat $\theta_0$/Ste. The calculated curve for nickel with no superheat closely follows that of aluminum for $\theta_0 = 0.3$ Ste. $T_0$ is the initial temperature of the droplets (from Reference 1).

The dimensionless interface velocity for Newtonian solidification is:

$$\left.\frac{r_0}{\alpha_\ell} R\right|_{\text{Newtonian}} = \left(\frac{1}{\phi^*}\right)^2 \text{Bi Ste} \frac{C_{p\ell}}{C_{ps}} \tag{5}$$

where  R = solid-liquid interface velocity
$\phi^*$ = $r^*/r_0$ = dimensionless solid-liquid interface position
$r^*$ = interface position

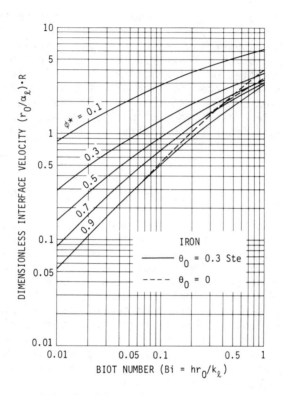

Figure 3.  Dimensionless solid-liquid interface velocity during solidification of iron droplets, as a function of Biot number and fractional position of the interface $\phi^* = r^*/r_0$. $\theta_0 = C_{ps}(T_0-T_M)/\Delta H_{s\ell}$ (from Reference 1).

For a given dimensionless interface position, $\phi^*$, equation (5) gives a straight line with a slope of unity in Figure 3. The ratio of calculated non-Newtonian to Newtonian interface velocities is close to unity in the beginning of solidification for the case of no superheat regardless of Biot number. On the other hand, this ratio decreases with increasing fraction solid and increasing Biot number at a given fraction solid. In general, increasing initial superheat in the non-Newtonian regime results in a corresponding decrease in solid-liquid interface velocities. However, this effect is only noticeable in the initial stages of solidification, and decays rapidly with increasing superheat and decreasing Biot number.

Note that during non-Newtonian solidification increasing the particle size by one order of magnitude (without changing h) reduces the actual interface velocity at a particular fraction solid. For example, assuming a heat transfer coefficient $h = 5 \times 10^4$ W/m$^2$K interface velocities of $1.1 \times 10^{-1}$ m/s and $5.3 \times 10^{-2}$ m/s are calculated for an interface position $\phi^* = 0.5$ in solidifying iron droplets of 50μm and 500μm, respectively. On the other hand, increasing the heat transfer coefficient h (without changing $r_0$) will increase the solid-liquid interface velocity, R, at any location.

## HEAT FLOW DURING SOLIDIFICATION AGAINST A METAL SUBSTRATE

In the recent past a large number of innovative batch and continuous techniques for production of rapidly solidified material against a metal substrate have been developed. Determination of exact cooling rates during solidification of crystalline and non-crystalline structures in these processes have required estimates of heat transfer coefficients between the melt and the substrate. Two measurements of heat transfer coefficient in splat cooling have been reported (3,4). Table II shows these values along with measured heat transfer coefficients for the case of a pressurized (13,000 psi) 89.7 MN/m$^2$ aluminum casting against a steel mold (5) and liquid aluminum die cast against a steel mold at 55 m/s metal flow velocity (6). Again, it appears than an upper limit exists for practically achievable heat transfer coefficients between liquid metals and substrates. It is probably in the range of $h = 10^5$ to $10^6$ W/m$^2$.K.

Computer heat flow calculations carried out for solidification of a non-crystalline aluminum melt are shown in Figures 4 and 5. The data is presented in terms of the dimensionless variables

$$Bi = \frac{hL}{k_\ell} \quad \text{and} \quad Fo = \frac{\alpha_\ell t}{L^2} \quad (5)$$

where L is casting thickness.

There is no evidence that aluminum has been made amorphous by rapid solidification, however, the numbers generated in these calculations are useful in that they again present the upper limit of

TABLE II

MEASURED HEAT TRANSFER COEFFICIENTS FOR SOLIDIFICATION OF
ALUMINUM AGAINST A METAL SUBSTRATE

| TECHNIQUE | THICKNESS OF CASTING | $h$, $W/m^2 \cdot K$ |
|---|---|---|
| 1. SPLAT ON Ni SUBSTRATE (from reference 3) | 1 μm | $1.1 \times 10^5$ to $2.8 \times 10^5$ |
| 2. DROP SMASH ON Fe SUBSTRATE (from reference 4) | 150 μm | $1.7 \times 10^4 - 1.8 \times 10^5$ |
| 3. PRESSURE CAST IN STEEL MOLD (from reference 5) | $10^4$ μm | NO PRESSURE $3.1 \times 10^3$<br>89.7 MPa $3.3 \times 10^4$ |
| 4. DIE CAST IN STEEL MOLD (from reference 6) | $1.6 \times 10^3$ μm | $7.94 \times 10^4$<br>Metal flow velocity 55 m/s<br>Pressure 175 MPa |

achievable cooling rates — heat of fusion did not have to be removed during solidification.

As expected, for small Biot numbers temperature gradients in the melt are negligible, Figure 4 (a) and the linear portion of the curve in Figure 5. The maximum melt thickness for this region of the plot is $\sim 40\mu m$ for a heat transfer coefficient h = 4.18 x $10^5$ W/$m^2$.K (which is higher than those reported in Table II). At Biot numbers larger than $\sim 10$ resistance to heat flow is primarily within the aluminum melt (cooling is essentially ideal) and dimensionless average cooling rate is independent of Biot number — actual average cooling rate $(\partial T/\partial t)_{Avg}$ is inversely proportional to the square of melt thickness. Thus, for Biot numbers in the Newtonian and ideal cooling ranges the actual average cooling rate in the liquid increases by one and two orders of magnitude, respectively, as the melt thickness is decreased by one order of magnitude. Thus, as in the case of metal droplets, the most effective way to increase cooling rate is to decrease melt thickness.

Similar calculations have been reported for crystalline solidification of iron splats against a copper substrate (7). It was found that for Biot numbers less than $\sim 0.015$ and more than $\sim 30$, cooling was Newtonian and ideal, respectively.

Analytical expressions are available for the solid-liquid interface velocity during plane front solidification of a melt against a substrate for the cases of Newtonian and ideal cooling, equations (6) and (7), respectively.

$$R = \frac{h}{\rho_s \Delta H_{s\ell}} (T - T_g) \qquad (6)$$

$$R = \frac{2\gamma^2 \alpha_s}{S} \qquad (7)$$

where R = solid-liquid interface velocity
$\alpha_s$ = thermal diffusivity of the solid forming

$\rho_s$ = density of the solid forming

S = distance solidified

$\gamma$ = argument of the error function solution of the temperature distribution; it is determined from a characteristic equation which contains metal and substrate thermal constants.

Approximate analytical solutions are available for Biot numbers between these two extremes. However, numerical solutions such as those described above can be readily employed.

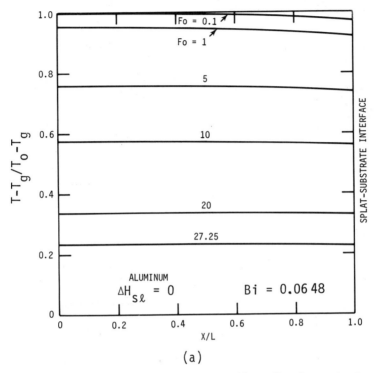

Figure 4. Calculated temperature distributions during cooling and non-crystalline solidification of splats of aluminum against a copper substrate. T, and $T_o$ and $T_g$ denote instantaneous and initial aluminum melt and initial copper substrate temperatures, respectively. X/L is fractional distance from the free surface of the aluminum. (from Reference 2).

Several attempts have been made to predict critical cooling rates in the liquid during RSP for the formation of non-crystalline structures (8-10). Heat flow calculations are usually combined with theories of nucleation, growth and transformation kinetics. This area, as well as, rapid solidification of undercooled crystalline structures remains fertile for future investigations.

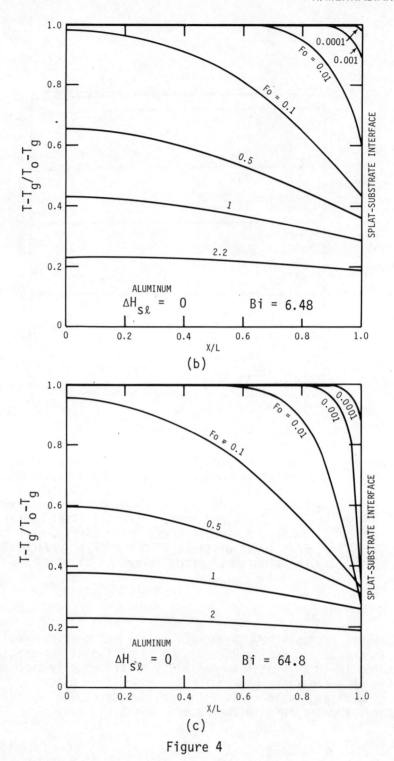

Figure 4

# RAPID SOLIDIFICATION PROCESSING

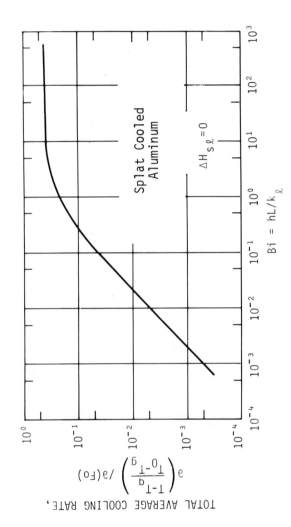

Figure 5. Dimensionless cooling rate averaged over melt thickness and time for temperature to reach half the melting point for non-crystalline solidification of an aluminum melt against a copper substrate. $T_0$ and $T_g$ are initial melt and substrate temperatures, respectively.

## HEAT FLOW DURING MELTING AND SOLIDIFICATION OF A SURFACE LAYER

The recent availability of high power directed energy sources such as the electron beam and the different types of lasers has led to the development of rapid melting and solidification techniques in which a bulk (semi-infinite) substrate in intimate contact with a molten layer acts as the quenching medium. Results of recent one and two-dimensional computer heat flow analyses (11,12) carried out to investigate the effect of high intensity radiation on the important surface layer melting and subsequent solidification variables of three substrate materials: aluminum, iron and nickel are summarized below.

The problem considered was the rapid melting and subsequent solidification of the surface layer of a semi-infinite solid, initially at room temperature, subjected to a high intensity stationary heat flux over a circular region on its bounding surface. Since the melt and the substrate are in intimate contact, heat transfer coefficient between the two tends to infinity. Thus, one important limitation on rapid rate of heat extraction encountered in atomization and splat cooling types of processes has been removed. However, as shown below, surface melting via directed energy sources has its own inherent limitation on maximum achievable cooling rate during subsequent solidification which is again a function of the thickness of the molten surface region.

In general, the absorbed heat flux distribution can be both a function of position within the circular region as well as time. The generalized, two-dimensional heat flow model and numerical solution techniques developed in oblate spheroidal coordinate system (12) can readily take into consideration both the space and the time variation of the absorbed heat flux. However, the numerical results presented below are for step function uniform and Gaussian heat flux distributions within the circular region on the bounding surface of the semi-infinite substrate. The relationship between these two absorbed heat flux distributions is shown in Figure 6. If we assume that the total power absorbed, Q, in the circular region is identical for the uniform and the Gaussian heat flux distributions, as shown in Figure 6, then the following relationship is readily deduced:

$$q_{uniform} = \frac{q_o}{2.313} \qquad (8)$$

where $q_o$ is the absorbed heat flux at the center of the circular region in the Gaussian distribution.

In what follows we will first consider one and two-dimensional heat flow for the case of uniform absorbed heat flux before treating the more complex Gaussian heat flux distribution.

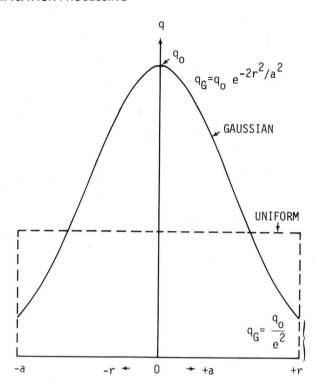

Figure 6. Relationship between uniform and Gaussian absorbed heat fluxes when the total power absorbed over the circular region is identical. q denotes the absorbed heat flux while a is the radius of the circular region. (from Reference 12).

Two important criteria were deduced, from the computer heat flow calculations (12), for the surface melting of a semi-infinite substrate subjected to a uniform heat flux q over a circular region of radius a on its bounding surface. First, it was shown that the product qa should exceed specific values, given by the following expression, if surface melting is to be initiated and the center of the circular region is to reach a given temperature up to the vaporization temperature of the substrate:

$$qa \geq k_\ell(T(0,0)-T_M) + \frac{k_s \Delta H_{s\ell}}{C_{p\ell}} + k_s(T_M-T_g) \qquad (9)$$

where $T(0,0)$ is the temperature at the center of the circular region, $T_g$ is the initial temperature of the substrate, $k_s$ is the thermal conductivity of the solid and all other terms have previously been defined.

The ratio of the two sides of equation (9) is plotted on the vertical axis on the right size of Figure 7. This plot, which was obtained from a large number of computer runs, shows that for a given radius of the circular region, $\underline{a}$, there is a minimum heat flux required if the center of the circular region is to reach a given temperature. That is for small values of $a/2\sqrt{\alpha_s t}$ the ratio on the vertical axis approaches one - the temperature $T(0,0)$ approaches its maximum steady state value. For an aluminum substrate, minimum values of $qa \simeq 2.3 \times 10^5$ W/m and $qa \simeq 4.2 \times 10^5$ W/m are deduced from expression (9) and Figure 7 for the initiation of surface melting and for a surface temperature $T(0,0) = T_v$, where $T_v$ denotes the vaporization temperature. For example, when $qa \simeq 2.86 \times 10^5$ the calculated maximum steady state temperature at the center of the molten zone of an aluminum substrate is 1607°K (12).

The second important criterion previously developed (12) was that for large values of the product $\underline{qa}$ isotherms in the substrate are planar in shape and one-dimensional heat flow conditions prevail - lateral heat flow in the substrate can be ignored. The vertical axis on the left side of Figure 7 is the ratio of the square root of time in one and two-dimensional heat flow. The curve associated with this axis shows that for values of the dimensionless term $a/2\sqrt{\alpha_s t} \gtrsim 2.0$ - the radius of the circular region is larger than the characteristic distance for diffusion of heat-heat flow is essentially one-dimensional. The specific value of $\underline{qa}$ deduced for a given temperature $T(0,0)$ for this condition to be met is 3.33 times the value calculated from expression (9). As example, for an aluminum substrate, when $qa \gtrsim 1.4 \times 10^6$ W/m heat flow is essentially in one-dimension for $T(0,0)$ temperatures up to the vaporization temperature. Time for the center of the circular region to reach any temperature above the melting point of the substrate can also be deduced from Figure 7. For example, for an absorbed uniform heat flux of $10^9$ W/m$^2$ and a spot radius $a = 380\mu m$, $T(0,0)$ will reach 1700K in $2.1 \times 10^{-3}$ sec. Furthermore, the maximum temperature achieved at the center of the circular region would be 2238K.

Results from one-dimensional heat flow analysis, when the criterion noted above is applicable, are shown in Figures 8 to 11. Calculated melt depths versus total time for several uniform absorbed heat fluxes are shown in Figure 8. The arrows in the figure denote times at which the surface of each material reaches it's vaporization temperature, $T_v$, and the heat flux, q, is removed. Melting continues a while longer until a maximum melt depth, $z_{max}$, is reached.

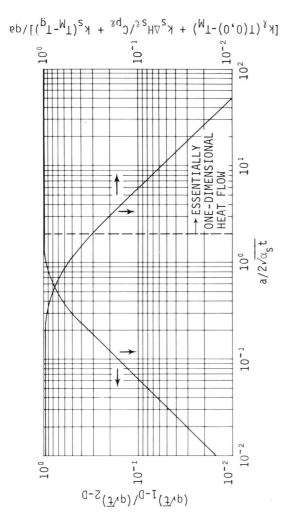

Figure 7. Temperature at the center of the liquid zone of a semi-infinite solid substrate during surface melting as a function of uniform absorbed heat flux, radius of the circular region and time. The vertical axis on the left gives the ratio of square root of time in one and two-dimensional heat flow (see Reference 12).

Note that for an aluminum substrate subjected to uniform absorbed heat fluxes of $5 \times 10^8$ W/m$^2$ and $5 \times 10^{10}$ W/m$^2$ the criterion for one-dimensional heat flow would be met if the radii of the circular regions exceed 2800μm and 28μm, respectively. The corresponding maximum melt depths, $z_{max}$, from Figure 8 are ~650μm and 6.5μm, respectively.

Melt depth during melting and subsequent crystalline solidification periods are expressed in dimensionless form, $\rho_s \Delta H_{s\ell} z / q t_m$ in Figure 9. $t_m$ denotes the time it takes for the surface of a material to reach its melting point and z is the melt depth perpendicular to the surface of the substrate.

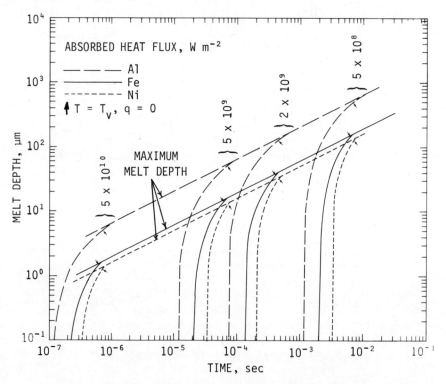

Figure 8. Melt depth versus total time for different uniform absorbed heat fluxes obtained via a numerical technique. Arrows indicate times at which a surface reaches it's vaporization temperature and heat flux is removed (from Reference 11).

# RAPID SOLIDIFICATION PROCESSING

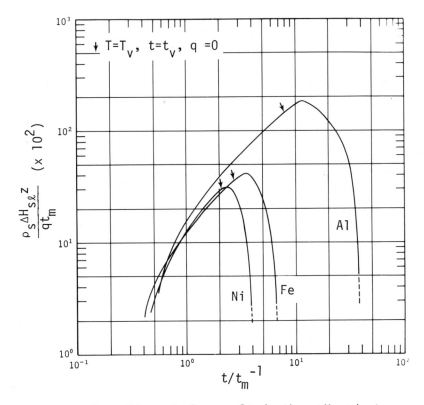

Figure 9. Dimensionless melt depth $\rho_s \Delta H_{s\ell} z/q\, t_m$ during melting and solidification versus dimensionless time, $t/t_m$ (from Reference 11).

Some general trends between the absorbed heat flux, time and melt depth can be readily deduced from the data presented in Figures 8 and 9.

The surface of a substrate reaches a given temperature when $q\sqrt{t}$ is kept constant. The significance of this finding is that at higher power inputs, less time is available for diffusion of heat into the metal substrate - the absorbed heat is concentrated near the surface of the material leading to steeper temperature gradients.

The melt depth is inversely proportional to the absorbed heat flux.

The change in temperature with time in the liquid, heating and cooling rates during melting and subsequent solidification, is given by the product $G_L \cdot R$ where $G_L$ is the temperature gradient and R is the liquid-solid interface velocity. The function $(G_L \cdot R)_{Avg}/q^2$ (instantaneous average cooling rate in the liquid phase at any time during crystalline solidification divided by the square of the absorbed heat flux) versus fractional distance solidified, $S/z_{max}$, is plotted in Figure 10. This cooling rate is a maximum at the beginning of solidification when temperature gradients in the liquid are high. Cooling rates during non-crystalline solidification were also calculated for the three substrate materials by setting the heat of fusion, $\Delta H_{s\ell}$, equal to zero during the solidification half of the numerical computations. These data, plotted in Figure 11, show that the ratio of total average cooling rate (cooling rate averaged over melt depth and time until the surface of the substrate reaches one-half the melting temperature) to the square of absorbed heat flux is a unique function of the fractional melt depth, $z/z_{max}$ where $z/z_{max}$ again denotes maximum melt depth achieved after the surface of the substrate reaches the vaporization temperature and the heat flux is removed. The significance of the data shown in Figures 10 and 11 is that average cooling rate in the liquid region and total average cooling rate during crystalline and non-crystalline solidification of a surface layer would increase by two orders of magnitude if the absorbed heat flux is increased by one order of magnitude. On the other hand, it was noted above that absorbed heat flux is inversely proportional to the melt depth. Therefore, as in other types of rapid solidification processing, cooling rates are inversely proportional to melt depth, i.e. the thickness of the rapidly solidified specimen.

For example, consider an aluminum substrate subjected to uniform absorbed heat fluxes of $5 \times 10^8$ W/m$^2$ and $5 \times 10^{10}$ W/m$^2$. As before, the corresponding maximum calculated melt depths are 650μm and 6.5μm, respectively. Now, assume that the heat fluxes are removed once 10% of these melt depths are achieved in each case, $z/z_{max} = 0.1$ in Figure 11, z = 65μm and 0.65μm, respectively. The corresponding total average cooling rates from Figure 11 for the lower and the higher absorbed heat fluxes are $\sim 1.9 \times 10^5$ K/sec and $\sim 1.9 \times 10^9$ K/sec, respectively. For comparison, assume a high heat transfer coefficient of $h = 10^5$ W/m$^2$K between an aluminum splat and a copper substrate. Using the thermophysical data for aluminum given in Reference (11) cooling rates of $\sim 2.3 \times 10^5$ K/sec and $\sim 2.3 \times 10^7$ K/sec are calculated from Figure 5 for splat thicknesses of 65μm and 0.65μm, respectively. Comparison of these values with those calculated above for the rapidly solidified surface layers reveals that the latter permits higher cooling rates as the melt depths are reduced by application of higher heat fluxes. This is due to the fact that in surface melting average cooling rate is inversely proportional to the square of the melt depth (is directly proportional to the square of the absorbed heat flux) while in splat cooling type processes, when solidification is in the Newtonian regime, average cooling rate is only inversely proportional to the melt thickness.

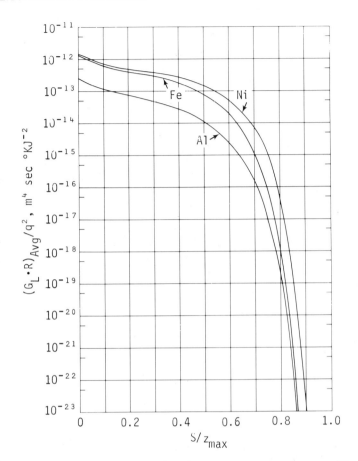

Figure 10. The ratio of instantaneous average cooling rate in the liquid to the square of absorbed heat flux, $(G_L \cdot R)_{Avg}/q^2$ versus fractional distance solidified, $S/z_{max}$ (from Reference 11).

The general trends established in these one-dimensional computer calculations (11) are summarized in Table III below. They show that temperature gradients in the liquid and solid phases and interface velocities are directly proportional to the absorbed heat flux, whereas melt depth is inversely proportional to the absorbed heat flux. Average cooling rates comparable to and exceeding those predicted for splat cooling can be achieved by increasing the heat flux and reducing the dwell time of the incident radiation. An order of magnitude increase in the absorbed heat flux results in a corresponding two orders of magnitude increase in average cooling rates in the liquid during solidification of crystalline and non-crystalline structures.

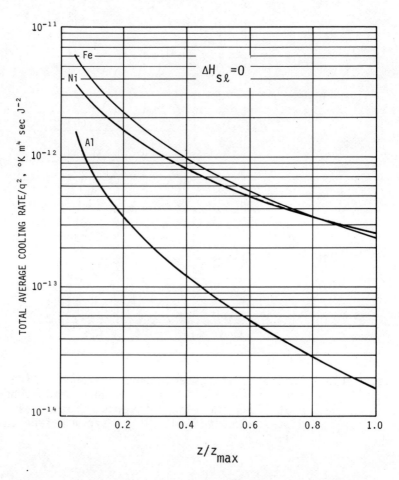

Figure 11. The ratio of total average cooling rate to the square of absorbed heat flux versus fractional distance melted, $z/z_{max}$, during solidification of a non-crystalline solid (from Reference 11).

## TABLE III

### One-Dimensional Heat Flow
### The Effect of Change in Absorbed Heat Flux on Other Variables

Order of magnitude increase in $\underline{q}$ results in the following changes:

| $\underline{q}$ | Time | Melt Depth | $\underline{R}$ | $\underline{G_L}$ | $\underline{G_L \cdot R}$ | $\underline{G_L/R}$ |
|---|---|---|---|---|---|---|
| 10 ↑ | $10^2$ ↓ | 10 ↓ | 10 ↑ | 10 ↑ | $10^2$ ↑ | ↔ |

---

↑ increase

↓ decrease

↔ no change

From Reference 11.

Results from the two-dimensional heat flow calculations (12) for a uniform absorbed heat flux q over a circular region of radius a on the bounding surface of a semi-infinite aluminum substrate are shown in Figures 12 and 13.

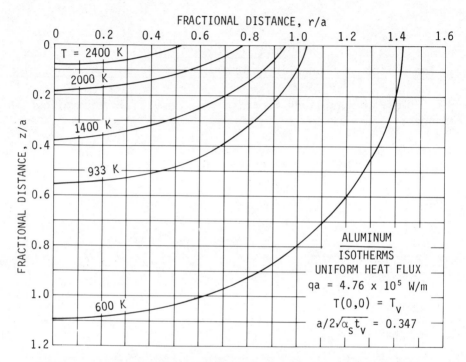

Figure 12. Shape and location of several isotherms, including the liquid-solid interface (T=933K), during melting of an aluminum substrate subjected to a uniform absorbed heat flux q over a circular region of radius a. The isotherms are drawn at the instant when $T(0,0)=T_v=2723K$ (from Reference 12).

An important finding from this work is that for a given value of T(0,0) at the center of the circular region the dimensionless temperature distribution in the substrate material during melting and solidification is identical for any combination q and a as long as qa = constant. Figure 12 shows the shape and location of several isotherms, including the liquid-solid interface, in an aluminum substrate material subjected to a uniform heat flux q over a circular region of radius a where the product qa=4.76 x $10^5$ W/m. Note that

RAPID SOLIDIFICATION PROCESSING                                      39

these are the isotherms at the instant T(0,0) reaches the vaporization temperature of aluminum $T_v$=2723K and the axis used are made dimensionless dividing them by the radius of the circular region. Identical dimensionless plots were obtained from several computer runs where q and a values were varied over large ranges while the product qa was kept constant.

Figure 13 shows the effect of increasing the product qa on the geometry and the location of the liquid-solid interface. The curve for qa ≃ 1.9 x 10⁶ W/m is in the range where heat flow can be assumed to be essentially one-dimensional and the shape of the isotherm verifies this fact. On the other hand, with decreasing values of qa, lateral heat flow becomes significant and the liquid-solid interfaces assume increasingly more convex geometries.

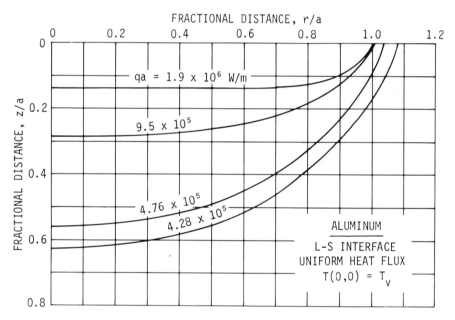

Figure 13. Liquid-solid interfaces during melting of an aluminum substrate subjected to a uniform heat flux q over a circular region of radius a. The interfaces for each constant qa are shown at the instant the center of the circular region reaches the vaporization temperature. (from Reference 12).

Using the relationships between uniform and Gaussian heat flux distributions developed in Figure 6 and expression (8) the following correlations between melt depths and shapes of liquid-solid interfaces were obtained (12). For a given temperature at the center of the circular region $T(0,0)$ and total power absorbed, $Q$, melt depths in the Gaussian heat flux distributions are $\sim 2.313$ times shallower than that obtained in the case of the uniform absorbed heat flux.

Figure 14 compares the shapes and locations of the liquid-solid interfaces between uniform and Gaussian heat flux distributions when $T(0,0)$ reaches the vaporization temperature of an aluminum substrate. As indicated above, for a given total absorbed heat flux expression (8) holds and the center of the circular region heats up much more rapidly when the heat flux has a Gaussian distribution. The melt depth along the z-axis is shallower, while the melt width is smaller due to the diminishing absorbed heat flux with increasing r in the circular region.

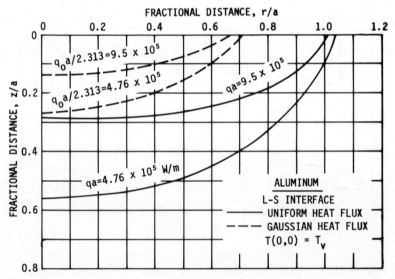

Figure 14. Comparison between the shapes and locations of the liquid-solid interfaces during melting of an aluminum substrate subjected to uniform and Gaussian absorbed heat flux distributions at the instant $T(0,0) = T_v$ (from Reference 12).

# RAPID SOLIDIFICATION PROCESSING

Figure 15 shows the general relationship developed from numerical computations between temperature at the center of the circular region and the product of absorbed heat flux and melt depth for an aluminum substrate. z in this figure denotes melt depth along the z-axis at the center of the circular region r = 0. For example, for

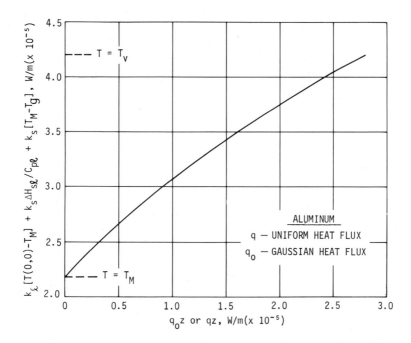

Figure 15. Temperature at the center of the circular region $T(0,0)$ over which a heat flux is absorbed in an aluminum substrate versus the product of the absorbed heat flux and melt depth in the axial direction away from the surface. $q_0$ denotes the maximum heat flux at the center of the Gaussian heat flux distribution. See Figure 6 and equation (8) (from Reference 12).

uniform heat fluxes of $3 \times 10^8$ W/m$^2$ and $2.5 \times 10^9$ W/m$^2$ absorbed over circular regions with radii of $\approx$ 1590μm and $\approx$ 190μm, respectively (qa $\approx 4.76 \times 10^5$ W/m) the corresponding times and melt depths during melting of an aluminum substrate when $T(0,0) = T_v$ are $t = 5.4 \times 10^{-2}$ sec and $t = 7.7 \times 10^{-4}$ sec from Figure 7 and z = 933μm and z = 112μm from Figure 15, respectively. Similar information can also be obtained from these two figures for any $T(0,0)$ temperature between $T_M$ and $T_v$. While the data plotted in Figure 7 can be used for any substrate material, the curve in Figure 15 is specifically calculated for an aluminum substrate. Similar curves can readily be generated for other substrate materials using the equations and computations procedure outlined in the previous sections. As noted above, the data in Figure 15 also show that for given total powers absorbed, Q, melt depths for the Gaussian heat flux distribution are 2.313 times shallower.

Figures 12 to 15 indicate that some of the general relationships developed between absorbed heat flux, melt depth and time in the one-dimensional heat flow model may be extended to the case of two-dimensional transient heat flow. The data also show that for a given temperature $T(0,0)$ at the center of the circular region and total power absorbed, Q, shallower melt depths and heat affected zones, higher temperature gradients and hence, higher cooling rates during solidification are achieved with the Gaussian heat flux distribution.

## SUMMARY

There are specific heat flow limitiations on the maximum achievable cooling rates in different Rapid Solidification Processing, RSP, techniques. In atomization and splat cooling types of processes the limitation on the rate of heat extraction is imposed by the practically achievable heat transfer coefficients. Once heat transfer coefficients in these processes are maximized, $h = 10^5$ to $10^6$ W/m$^2$K, further improvements in cooling rates can only be obtained at the expense of reduced specimen size or thickness. The rapid surface melting and solidification technique, via directed high energy sources, has its own limitation on specimen size and thickness, even though the melt is intimate contact with its own substrate. High cooling rates during solidification are only achieved at the expense of high absorbed heat fluxes, reduced heat affected zones and melt depths. Furthermore, the requirement of high absorbed heat flux imposes restrictions on the spot size (e.g. the radius) of the surface melted region.

## REFERENCES

1. C. G. Levi, R. Mehrabian, submitted for publication to Met. Trans.

2. R. Mehrabian; Proceedings of Conference on Rapid Solidification Processing, Principles and Technologies, Nov. 1977, Reston, Virginia, p. 7. R. Mehrabian, B. H. Kear and M. Cohen, Eds., Claitor Publishing Division, 1978.

3.  P. Predecki, A. W. Mellendore and N. J. Grant, Trans. Met. Soc. AIME, 1969, Vol. 233, p. 1581.

4.  D. R. Harbur, J. W. Anderson and W. J. Maraman, Trans. Met. Soc. AIME, 1969, Vol. 245, p. 1055.

5.  S. D. E. Ramati, G. J. Abbaschian, D. G. Backman and R. Mehrabian, Met. Trans. B, 1978, Vol. 9B, p. 279.

6.  S. Hong, D. G. Backman and R. Mehrabian, submitted for publication to Met. Trans.

7.  R. C. Ruhl, Mater. Sci. Eng., 1967, Vol. 1, p. 313.

8.  P. H. Shingu and R. Ozaki, Met. Trans. A, 1975, Vol. 6A, p. 33.

9.  D. R. Uhlmann, J. Non-Cryst. Solids, 1976, Vol. 7, p. 337.

10. F. Spaepen and D. Turnbull; Proceedings of Second International Conference on Rapidly Quenched Metals. Edited by N. J. Grant and B. L. Giessen, 1975, M.I.T. Press, Cambridge, Mass., p. 205.

11. S. C. Hsu, S. Chakravorty and R. Mehrabian, Met. Trans. B, 1978, Vol. 9B, p. 221.

12. S. C. Hsu, S. Kou and R. Mehrabian, submitted for publication to Met. Trans.


ACKNOWLEDGEMENT

This research is being sponsored by the Defense Advanced Research Projects Agency and monitored by the Office of Naval Research under contract # N00014-78-C-0275. Technical monitor of the contract is Dr. B. A. MacDonald.

LASER PROCESSING OF MATERIALS

B. H. Kear, E. M. Breinan, & E. R. Thompson

United Technologies Research Center
East Hartford, Connecticut 06108

ABSTRACT

This paper presents an overview of developments, old and new, in the technology of laser processing of materials including transformation hardening and surface alloying, welding, cutting, drilling, rapid solidification processing, pulse annealing, shock hardening, laser assisted machining, and laser controlled surface reactions. Several current industrial applications are described. It is concluded that the unique capabilities of laser systems will lead to the realization of the laser as a commercially important materials processing tool.

INTRODUCTION

The adaptation of lasers to materials processing tasks got its start in the mid-1960's. At that time, since only pulsed lasers were available commercially, applications were limited to such processing operations as hole drilling, trimming and spot welding (1,2). With the advent of continuous wave lasers the emphasis shifted towards the development of cutting (3-5), welding (6-13), and heat treating (14-17) operations. There are now a number of industrial cutting and heat treating applications using high power lasers, as well as a few significant welding applications.

Within the past few years, successful demonstrations have been made of the usefulness of lasers in other areas of materials processing; namely, shock hardening, (18, 19), rapid solidification processing (20-22), pulse annealing (23, 24), and laser-assisted machining (25). Although no commercial applications have as yet been realized, the potential is clearly there, so that it is only a matter of time before these processes find their way into industry. At the present time, laser processing research is expanding rapidly into new areas such as atomization and controlled surface reactions. The future therefore, appears to hold great promise for a rising tide of successful applications for lasers in material processing tasks.

The purpose of this paper is to present an overview of both old and new developments in laser processing of materials and to describe commercial laser processing applications.

## STATE-OF-THE-ART

### Transformation Hardening and Surface Alloying

Fig. 1 defines the operational regimes of lasers for various industrial materials processing techniques. As indicated, laser-induced transformation hardening employs relatively low power densities of $10^3$-$10^4$ watts/cm$^2$, with interaction times of 0.01 - 1.0 secs.

In this process, the material surface is hardened by exploiting some naturally occurring solid state phase transformation, such as a martensitic transformation. To obtain the desired effect, the material is laser heated to some pre-determined surface temperature, and allowed to undergo rapid self-quenching. The depth of the hardened zone can be related to the incident power density, the interaction (dwell) time, and the temperature at the surface of the material, Fig. 2. Thus, knowing the depth of hardened zone desired, it is a simple matter to select the power density and interaction time in order to determine the beam parameters (i.e., laser power and spot diameter) required for a given processing rate. In practice, this means that the workpiece is usually located at some distance from the exact focal spot of the beam, where the energy density is appropriate. In order to obtain a specific surface coverage of hardened material, the beam is mirror-scanned over the surface of the workpiece, or the workpiece itself is indexed with respect to the beam. Overlapping laser hardened zones invariably show some indications of tempering in the regions of overlap, but the effect is not usually large. Table I shows how the depth of hardening and surface coverage rate depend on laser processing parameters in cast-iron. Figure 3 shows a typical laser hardened zone in this material. The material is hardened by a martensitic transformation.

# LASER PROCESSING OF MATERIALS

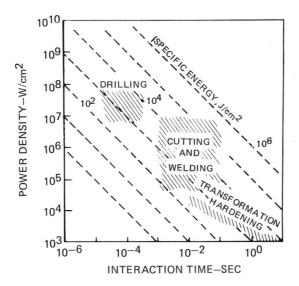

Fig. 1  Operational regimes for industrial laser processing.

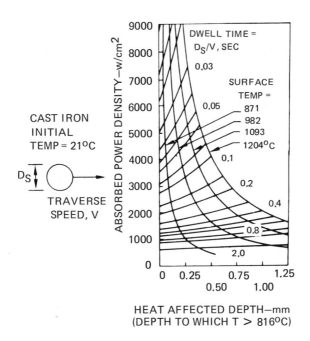

Fig. 2  Laser heat treating parameters.

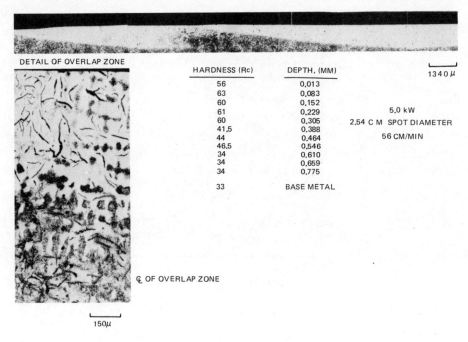

Fig. 3 Cross section of transformation hardened grey cast iron.

Table I

Laser Hardening Performance Summary

| Spot dia -cm | Traverse speed -cm/sec | Power -kW | Max depth -cm | Effective absorptivity | Max width -cm | Coverage rate -cm2/min |
|---|---|---|---|---|---|---|
| 0.63 | 10.60 | 2.5 | 0.036 | 0.63 | 0.62 | 394 |
| 1.27 | 2.54 | 2.5 | 0.048 | 0.74 | 1.01 | 153 |
| 1.27 | 10.60 | 5.0 | 0.036 | 0.81 | 1.02 | 645 |
| 1.80 | 3.18 | 5.0 | 0.052 | 0.65 | 1.43 | 271 |

For the purpose of applying a separate alloy to a materials surface (surface alloying), provision must be made to pre-place or deliver controlled amounts of the coating material to the interaction zone as the beam traverses over the surface of the workpiece. A successful demonstration of preferential hardening by an infused layer is reported for the case of tool steel coated by silk-screened, powdered tungsten carbide. Scanning a pulsed laser fused the carbide powder into the alloy steel surface producing a layer with a hardness 2.5 times that of the base tool. Such a method provides a simple approach for selective hardening.

## Deep Penetration Welding

Laser welding is achieved using a beam focussed at or near its minimum spot diameter, with power densities of $10^5$–$10^7$ watts/cm$^2$ and interaction times of 0.001–0.1 secs, Fig. 1. Under these conditions, laser energy is delivered to the workpiece surface more rapidly than it can be dissipated by conduction and radiation from the incident spot. The effect is to generate a vapor column, or deep penetration cavity, that extends through the workpiece. With appropriate motion of the workpiece relative to the focussed beam, the cavity can be translated across the workpiece to produce a deep penetration fusion weld, Fig. 4. Laser welds typically exhibit a high depth-to-width ratio, Fig. 5, which is the result of the extreme localization of energy deposition in the penetration cavity. Deep penetration autogenous welds have been successfully produced in a large number of materials, including low carbon steels, stainless steels, nuclear reactor materials, titanium alloys and superalloys. In general, the welds have exhibited excellent mechanical properties.

When a high power density laser beam is employed for deep penetration autogenous welding, the fusion zone exhibits not only a general refinement in microstructure, but also a marked reduction in visible inclusion particles, Fig. 6, as well as in soluble gas content. The effect which we have termed fusion zone purification is especially important in materials which contain a substantial quantity of inclusions. In X-80 and HY-130 alloy steels, the microstructural refinement and purification has provided laser weld zones with impact resistance substantially above that of the base alloys.

## High Speed Cutting

Cutting is performed with the same laser beams used for welding. The sharply focussed beam melts and vaporizes the material along the cut seam, and a high pressure gas jet expels the melt from the interaction zone. When reactive materials, such as titanium alloys, are

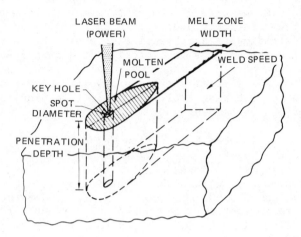

Fig. 4  Schematic of deep-penetration laser welding.

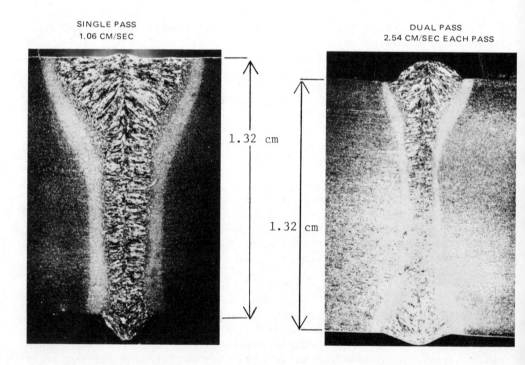

Fig. 5  Laser welds in X-80 Alloy.

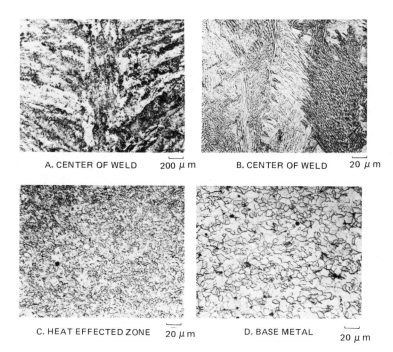

Fig. 6 Microstructures of laser welded X-80 illustrating fusion zone purification.

being cut, exothermic reactions involving the melt and constituents in the gas jet can be exploited to reduce the laser energy needed for cutting. Gas jet requirements can also be reduced if the material being cut has a high vapor pressure, and readily vaporizes out of the interaction zone. Laser cutting has been demonstrated for material up to 5.0 cm in thickness, Table II. However, the quality of cut deteriorates with increasing thickness, owing to the low depth of field in the laser focal spot. Considerable internal sidewall reflection of laser energy is necessary to produce cuts with even approximately parallel edges. The most promising applications, therefore, are those in which high speed cutting is performed on thin sections that are difficult to cut mechanically, such as composites and steel-reinforced hydraulic hose. In the cutting of thin metal sheet, narrow cuts with very small heat-affected zones can be produced at high rates. Figure 7 shows examples of cuts in Waspaloy sheet, titanium honeycomb panels, and 6061 aluminum sheet.

Table II

Jet-Assisted Laser Cutting Performance

| Material | Thickness cm | Cut speed, cm/min. | Power, kW |
|---|---|---|---|
| Aluminum | 0.10 | 635 | 3.0 |
|  | 0.32 | 254 | 3.0 |
|  | 0.64 | 102 | 3.0 |
|  | 1.27 | 76 | 3.0 |
| Nickel alloy | 0.32 | 305 | 4.0 |
| Stainless steel | 0.32 | 254 | 3.0 |
| Steel | 0.32 | 406 | 4.0 |
|  | 1.68 | 114 | 4.0 |
|  | 5.40 | 33 | 15.0 |
| Titanium | 0.16 | 558 | 0.25* |
|  | 0.64 | 304 | 0.25* |
|  | 1.27 | 152 | 0.25* |

*Oxygen assist

WASPALOY SHEET
LASER POWER: 3 kW
CUTTING SPEED: 14.82 cm/sec
THICKNESS: 0.158 cm

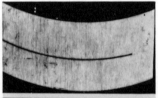

TITANIUM HONEYCOMB PANEL
LASER POWER: 3 kW
CUTTING SPEED: 3.39 cm/sec
THICKNESS: 0.635 cm

ALUMINUM SHEET
LASER POWER: 3 kW
CUTTING SPEED: 4.23 cm/sec
THICKNESS: 0.318 cm

Fig. 7  Laser cuts in several aerospace materials.

LASER PROCESSING OF MATERIALS 53

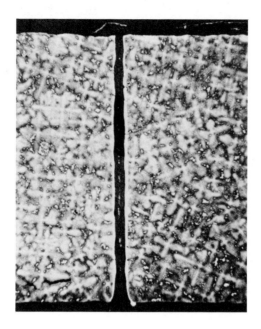

Fig. 8  Laser drilled hole in B-1900

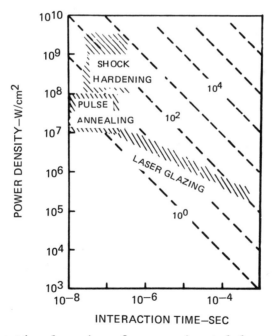

Fig. 9  Operational regimes for experimental laser processing.

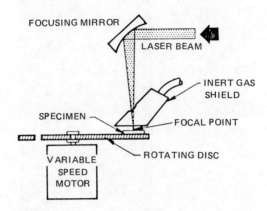

Fig. 10  Schematic of apparatus for rapid laser surface melting and self-substrate quenching.

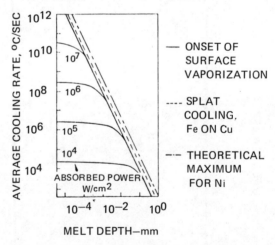

Fig. 11  Effect of melt depth and absorbed power on average quench rate.

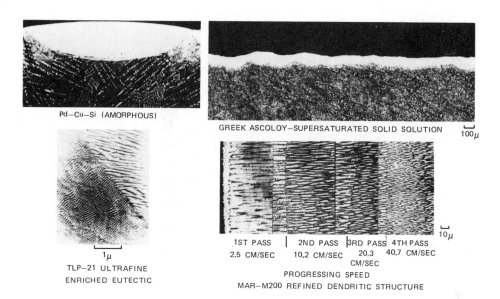

Fig. 12 Microstructures of Rapid Laser Surface Melted and Self-Substrate Quenched Alloys.

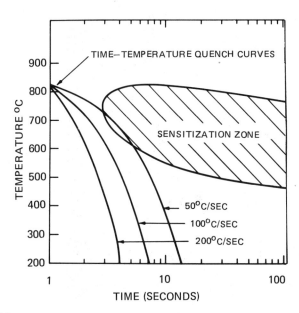

Fig. 13 Sensitization diagram for 304 Stainless Steel (After Ref. 22).

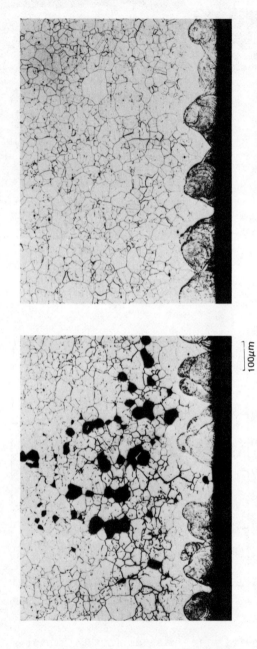

Fig. 14 Corrosion susceptibility of sensitized 304 stainless steel with (A) partial and (B) complete overlapping laser melt zones. (After Ref. 22).

LASER PROCESSING OF MATERIALS

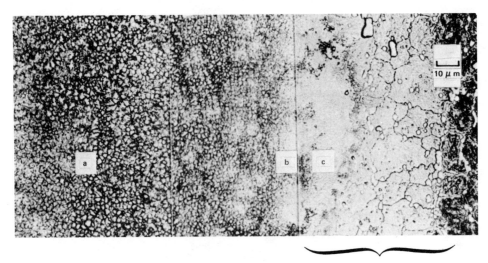

Fig. 15  Laser melted M2 Tool Steel
    (a) Duplex structure of δ-ferrite and austenite.
    (b) Austenite and trace of δ-ferrite.
    (c) Fully transformed austenite.
       (After Ref. 31)

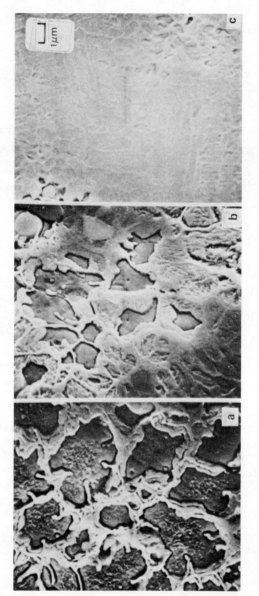

Fig. 16  Scanning Electron Micrographs of Laser Melted M2 Tool Steel
(a) Evidence of peritectic δ-ferrite to austenite transformation (outer melt zone).
(b) Intermediate melt zone showing some untransformed δ-ferrite.
(c) Near melt/substrate interface showing fully transformed austenite (After Ref. 31).

## Hole Drilling

Drilling of holes with lasers requires high power densities ~$10^7$ watts/cm$^2$, Fig. 1. The cavity is produced by an explosive vaporization effect, which expels material from the hole partially as liquid and partially as vapor. Holes with 35:1 aspect ratio can readily be produced in metallic materials. A hole drilled in a superalloy by a pulsed laser is shown in Fig. 8. The ability of the laser to drill holes at a small angle of incidence to the workpiece surface is a useful feature. This has been exploited, for example, in the drilling of deep, slanting holes in the platforms of gas-turbine blades.

## CURRENT DEVELOPMENTS

### Surface Melting

Surface localized melting, followed by rapid solidification and subsequent solid state cooling, is a new method for the surface treatment of materials; we have termed this the laserglaze$^{TM}$ process. The laser parameters required for this process are shown in Fig. 9. A high power density beam, similar to that employed in welding, is required together with an interaction time on the order of $10^{-4}$ - $10^{-6}$ secs. Processing is accomplished by rapidly traversing a sharply focussed laser beam over the material surface as depicted in Fig. 10. Under the appropriate conditions, surface melting is achieved with high melting efficiency, i.e., most of the energy absorbed causes melting and very little goes into heating the substrate. It is this feature that is responsible for causing rapid solidification, after cessation of laser energy input. Calculations of average cooling rates show a strong dependence on melt depth and absorbed power density, Fig. 11. As indicated, cooling rates up to $5 \times 10^6$ °C/sec are possible in $2.5 \times 10^{-2}$ mm thick melt layers in nickel. A variety of metallurgical microstructures have been produced by laserglazing, some of which are unique. These include amorphous metallic solids, extended (supersaturated) solid solution phases, metastable phases, ultrafine eutectics, and refined dendritic structures, Fig. 12.

Potential applications are currently being evaluated. An interesting case is the use of laser surface melting to combat stress corrosion cracking in 'sensitized' 304 stainless steel (22). This is achieved by adjusting the cooling rate to avoid the formation of a harmful carbide phase at the grain boundaries, Fig. 13. The effectiveness of this surface treatment in providing resistance to chemical attack is dramatically shown in Fig. 14, which compares the effect of partial and complete surface coverage of overlapping laser melt zones.

Another example is surface hardening of M-2 high speed steel by rapid solidification, followed by heat treatment (31). An optimum hardening treatment appears to be laser melt quenching at rates in excess of $10^4$°C/sec to obtain a supersaturated δ-ferrite/γ-austenite structure, Figs. 15 and 16 followed by aging at 1230°C to generate a uniformly fine dispersion of MC carbide phase, Fig. 17. The fine dispersion of carbide particles gives a peak hardness of VHN ~1100, which compares with VHN ~ 850 for conventionally processed material. It remains to determine how much influence size and distribution of the carbide phase, at a fixed volume fraction, has on the performance characteristics of the tool steel. A further example is laser melting of commercially fabricated Zircaloy-4 to improve its resistance to stress corrosion cracking (32). In this case, cooling rates of ~$10^5$°C/sec generate a very fine hcp martensitic structure, accompanied by the complete suppression of precipitation of intermetallic phases, Fig. 18. Prior experience with titanium-base alloys leads one to anticipate that such a fine martensitic structure will possess improved resistance to stress corrosion cracking.

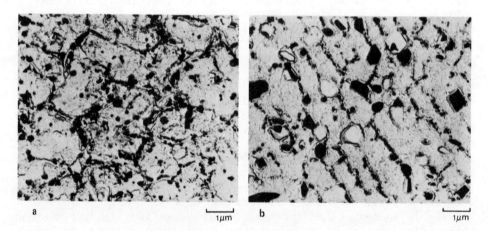

Fig. 17 Carbide extraction replicas from thermally treated, laser melted M2 tool steel
 (a) 560°C, 2 hrs. - air cooled
 (b) 1230°C, 5 mins. - quenched in liquid nitrogen
     (After Ref. 31).

## Bulk Rapid Solidification

The primary limitation or rapid surface melting and self-substrate quenching, as well as all other known techniques of rapid solidification processing, is that the product is in the form of a thin section. A concept, which we have termed the layerglaze process, was formulated to overcome this limitation and make possible the fabrication of bulk rapidly solidified structures. Fig. 19 shows a schematic of the process and a section of stainless steel prepared by this process. As indicated in the schematic, bulk rapidly solidified structure is built-up incrementally on a rotating mandrel, simply by laserglazing one thin layer of feedstock directly upon another in a continuous manner. Although the feedstock is shown being delivered to the mandrel in wire form, it is clear that a metered flow of melt from a tundish would serve just as well, and with less of a demand on laser energy required to effect the glazing operation. Using a continuous melt feed, the process of layerglazing is reminiscent of the familiar process of melt spinning for producing rapidly solidified filamentary material. The important difference being, of course, that in melt spinning the solidified filament is allowed to detach from the rotating wheel, whereas in layerglazing the deposited material is fused to the wheel. The layerglaze process appears to be particularly well suited to the fabrication of parts having an axis of rotational symmetry. The process is being considered for making discs.

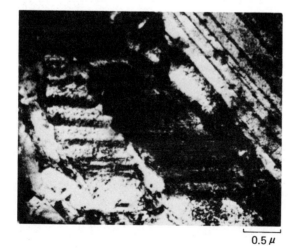

Fig. 18  Zircaloy$^{-4}$ glazed at 6 kW and 89 cm/sec (After Ref. 32).

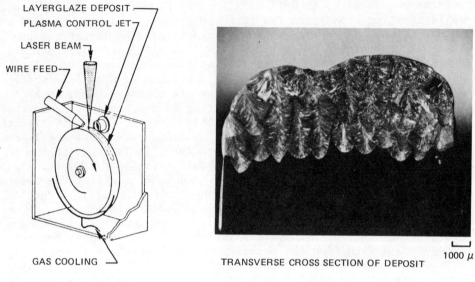

Fig. 19  Schematic of Layerglaze process and section of stainless steel deposit on periphery of disc.

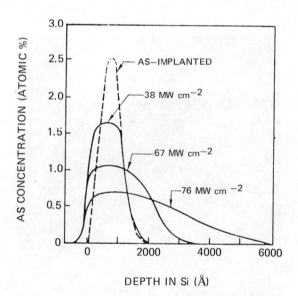

Fig. 20  Concentration profiles in pulse-annealed ion implanted silicon (After Ref. 24).

LASER PROCESSING OF MATERIALS 63

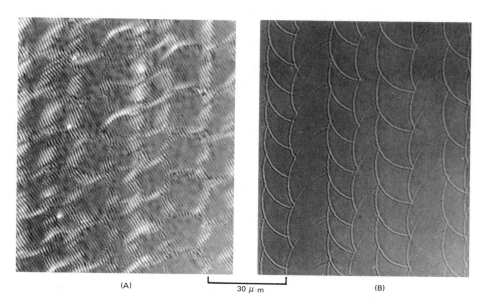

Fig. 21  Nomarski interference micrographs of pulse-annealed ion implanted silicon (a) 48 MW/cm$^2$, (b) 76 MW/cm$^2$ (After Ref. 24).

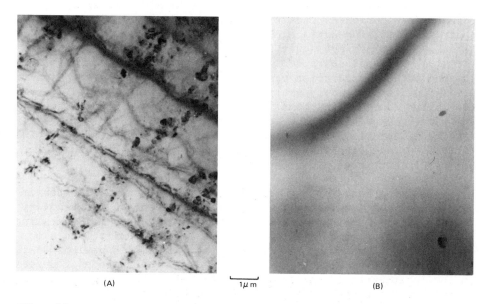

Fig. 22  Transmission electron micrographs of pulse-annealed ion implanted silicon (a) 48 MW/cm$^2$ (b) 76 MW/cm$^2$ (After Ref. 24).

The full potential of the layerglaze concept will only be realized when alloys have been designed properly to benefit from the initial rapid solidification. It appears that this can be done most simply by generating extended, or supersaturated solid solution phases by rapid solidification, and following this by phase decomposition under controlled conditions, so as to obtain the optimum distribution of hardening phase.

## Pulse Annealing

Ion implantation is commonly employed in the semi-conductor industry to introduce controlled amounts of pure dopants into the surface regions of semi-conductor crystals. The act of implantation, however, causes extensive damage to the crystalline material such that the implanted surface layer assumes all the characteristics of an amorphous, or highly disordered solid. Recently, it has been shown that intense pulses (Fig. 9) of laser radiation can cause the disturbed surface layers to recover their original crystallinity, while at the same time maintaining a high degree of substitutionality for the implanted ions. (23)

Figure 20 shows the influence of power density on the distribution of As dopant in a silicon substrate, after irradiation with a Nd-YAG laser (24). As shown, the depth distribution of implanted ions can readily be controlled by varying the power of the laser pulse. With power densities in the range $5-8 \times 10^7$ watts/cm$^2$, the dopant distribution extends to the free surface of the material, in contrast to the sub-surface enrichment typical of as-implanted material. This range of power densities also effectively restores the single crystal character of the silicon. Thus, spatially overlapping pulsed irradiation generates a continuous single crystal layer from the initially amorphous structure, Figs. 21 and 22, and the implanted ions exhibit a high degree of substitutionality. Lower power densities induce only partial recovery of the crystal lattice. Typically, the perimeters of the laser irradiated regions exhibit surface rippling, which correlates with the presence of polycrystalline regions in the annealed material. (25)

A similar behavior has been found in laser annealed boron implanted silicon (26). Figure 23 shows the extent of the recovery process due to laser annealing, as compared with thermal annealing. The thermally annealed material retains a high density of dislocations loops and stacking faults, whereas the laser annealed material is completely free of such lattice defects. The absence of defects in the laser annealed material is reflected in the complete recovery of the carrier concentration. Diffusion induced dislocation loops and precipitates can also be removed by high energy pulsed laser treatment. (27) Thus, boron or phosphorous atoms previously contained

in small precipitates become dispersed and electrically active, and
the resulting dopant concentration can exceed the normal solid solubility limit. Furthermore, deposits of boron on silicon at room
temperatures may be dissolved in the silicon by laser irradiation
and become electrically active, Figs. 24 and 25. This phenomenon
has been exploited to form p-n junctions by the simple expedient of
laser induced diffusion of a boron deposit into n-type phosphorous
doped silicon. (28) Solar cells with efficiencies of 10% have been
fabricated by this means. Since the technique does not involve any
thermal treatments at high temperatures, it should be faster and more
economical than the conventional diffusion technique.

## Shock Hardening

At very high power densities of $\sim 10^9$ watts/cm$^2$, nearly instantaneous surface vaporization occurs on interaction of the laser
beam with the material. This is the regime of laser shock hardening,
Fig. 9. If pulse duration is kept to $\sim 10^{-8}$ secs, typically involving
an energy input of 20 J/cm$^2$, interaction is limited to the surface
and the rapid expansion of vaporized metal produces an effect similar
to a blast wave. As a consequence, a shock wave propagates and reflects within the material causing significant work hardening. The
effect has been demonstrated in several aluminum alloys, including
both solid-solution strengthened and age-hardenable alloys (19,33).

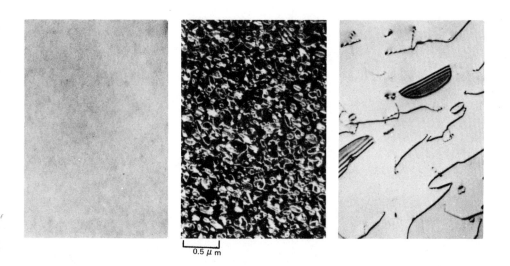

Fig. 23  Laser and thermally annealed B-implanted silicon
  (a) Laser - pulse energy $\sim$ 1.7J
  (b) Thermal - 900°C, 30 min.
  (c) Thermal - 1100°C, 30 min. (After Ref. 26).

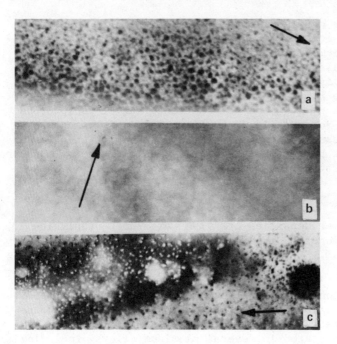

Fig. 24 - Transmission electron micrographs showing (a) as-deposited boron clusters on the silicon surface (b) absence of boron clusters in laser treated specimen and (c) re-precipitation of boron after annealing a laser treated specimen exceeding the solubility limit (Arrows represent 220 diffraction vectors and correspond to 0.15 µm) (After Ref. 28)

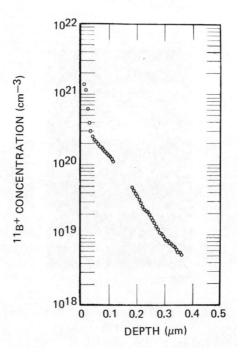

Fig. 25 Boron concentration after laser annealing of surface deposited boron. (After Ref. 28).

LASER PROCESSING OF MATERIALS 67

In particular, it has been shown that the laser shock hardening technique can be exploited to reharden weld and heat-affected zones in welded aluminum structures, Figs. 32 and 33. Although some surface melting normally occurs under such intense laser irradiation, this can be avoided by employing a sacrificial coating, such as black paint or a metal foil. (34) This raises the possibility of utilizing the shock hardening technique to improve the fatigue properties, as well as the strength of welded structures.

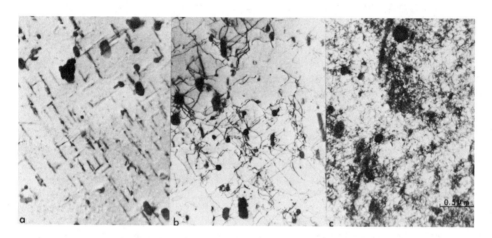

Fig. 32 Microstructure of 6061-T6 Aluminum Sheet Before and After Welding and laser shocking. (a) Initial microstructure - 6061-T6, (b) After welding-in the heat affected zone adjacent to the weld, (c) After laser shocking with a split beam, 25 ns pulse length and an energy density of 32.6 & 31.9 J/cm$^2$ on either surface - in the heat affected zone adjacent to the weld. (After Ref. 33)

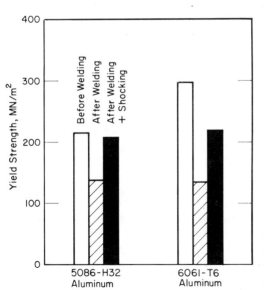

Fig. 33 Comparison of the 0.2% offset yield strength of welded and laser shocked aluminum alloys. (After Ref. 33)

## Laser Machining

In order to reduce the energy requirements for machining operations, and to increase cutting speeds, the concept of softening the workpiece just ahead of the tool bit had gained widespread acceptance. Currently, the most successful method for doing this employs heating by means of a plasma arc. Systems of this type have been adapted for high speed cutting of steels and other materials.

It is now recognized that lasers can be used for the same purpose, and perhaps with some advantage because of the ease with which the laser energy can be directed at a specific location on the workpiece. A study is currently underway (29) to demonstrate the effectiveness of a focussed laser in assisting the removal of material from a workpiece. To date, machining of a number of steels, superalloys, and ceramics has been accomplished. The configuration used for machining is shown in Fig. 26. A continuous 460W $CO_2$ laser beam, focussed to provide a maximum average power density of 3.5 $MW/cm^2$, is directed to the shoulder of the cut at a distance approximately 0.4 cm ahead of the tool. Through laser assistance, a machined surface with improved smoothness was generally obtained. Further, in certain cases, reductions in tool forces between 25 and 50% were measured. As shown in Fig. 27, smoother, more uniform cutting was also observed in the machining of Udimet 700. No tool wear was evident after the laser assisted removal of approximately 0.35 $cm^3$ of the nickel alloy while removal of a similar volume without laser assist caused the breakage of two ceramic tools.

A laser acting on the surfaces of alumina and silicon nitride has controllably removed material without the use of a cutting tool. Continuous spalling of the alumina surface has resulted in a finish similar to that obtained by conventional grinding. A square cross-sectional bar of silicon nitride was turned into a cylindrical rod by 'ablative machining' and the resultant rod was threaded using the focussed beam.

The use of a high power density laser beam as a means of assisting or producing the removal of material occurs with little heat leakage into the substrate thereby minimizing thermal distortion and thermal shock of the workpiece. Surface contouring of ceramics is an attractive potential application of laser machining.

## Laser Controlled Surface Reactions

When a laser beam is used to heat or melt thin layers of a substrate material in a chemically reactive rather than an inert gas atmosphere, constituents of the reactive gas may be deposited on the substrate, as in chemical vapor deposition (CVD), or they may be absorbed by the substrate. The general effect of surface gas absorp-

# LASER PROCESSING OF MATERIALS

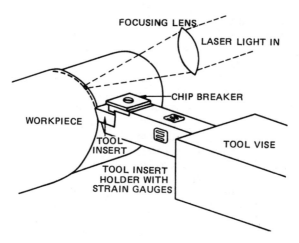

Fig. 26  Schematic of arrangement for laser assisted machining (After Ref. 29).

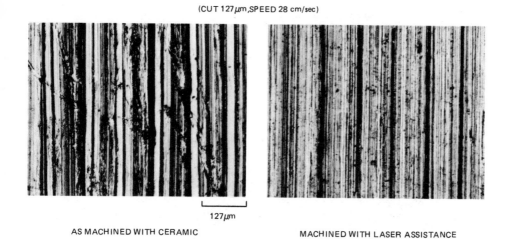

Fig. 27  Surfaces of Udiment 700 machined with ceramic tool (After Ref. 29).

tion will be to produce an alloy gradient from surface to bulk, with no sharp discontinuities in microstructure and mechanical properties. This could be useful in applications requiring coatings that are integrally bonded to the workpiece. Laser controlled CVD processing (30) also appears to be a promising technique for depositing new constituents or phases on substrate surfaces, Fig. 28. So far this process has been demonstrated only for the case of silicon deposition on quartz or sapphire substrates, via the pyrolysis of $SiH_4$, Fig. 29. However, it seems clear that the process can be applied with equal facility for the purpose of obtaining deposits of boron by reduction of $BCl_3$, alumina by hydrolysis of $AlCl_3$, silicon carbide by pyrolysis of $(CH_3)_2SiCl_2$, and tungsten or molybdenum by reduction of the chlorides. The ability of the process to give high spatial resolution in surface deposition, as has been shown for silicon deposition on quartz, may turn out to be its most valuable feature. It certainly gives credibility to the idea that the process may be employed to 'write' semiconductor, oxide, or conductor patterns on appropriate substrates in microelectronic components, or even to apply corrosion or wear resistant deposits at specific locations of components designed for structural applications.

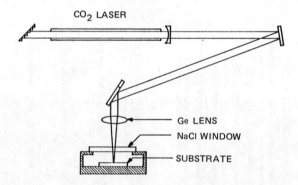

Fig. 28  Laser controlled chemical vapor deposition processing (After Ref. 30).

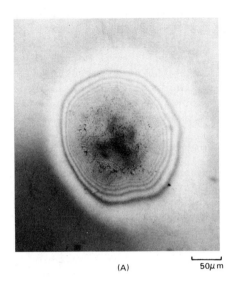 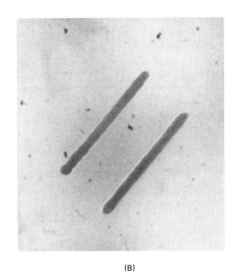

(A)   50μm   (B)

Fig. 29   Silicon Deposits on Quartz produced by laser CVD process
(a) isolated spot ~ 1.2 mm diam.
(b) overlapping spots - line width ~ 50 μm)
(After Ref. 30).

## APPLICATIONS

### Automotive

A wide variety of real and potential laser applications in the automotive industry have already been identified.  These range across the entire gamut of processing techniques from welding (e.g., gas tanks, underbodies and wheel rims), cutting (e.g., fabrics, glass), transformation hardening (e.g., valve stem guide holes, cylinder bore walls), hardfacing (e.g., valves, valve seats), and hole-drilling (e.g., fuel injector assemblies).  A typical example of a heat-treating application is the use of a laser beam to surface-harden the inner bore of valve stem guides, which has already shown a tremendous decrease in the wear rate of these holes.

Perhaps the best example of a complete laser system for an automotive process is the underbody welding system delivered by UTC to the Ford Motor Company in 1974.  This 5-axis, 6kW system is in intensive use at the Ford Manufacturing Research & Development Facility for the development of new high-power laser applications.

The largest production application of lasers to automotive

processing at the present time is by the Saginow Steering Gear Division of General Motors when transformation hardening is carried out on the inside bore of power steering gear assemblies, Fig. 30. A second large production application is to be in General Motors Electromotive Division where four, 5 kW systems will transformation harden the walls of a cast iron cylinder liner for a diesel locomotive engine.

## Aerospace

An important application in the aerospace industry, specifically at Grumman, is the cutting of titanium sheet. Exothermic reactions with oxygen permits sheet up to 1/2 in. in thickness to be cut by a 250 watt cw laser. Lasers are also extensively used in industry for the drilling of small diameter holes, particularly in hard to drill materials. The primary example of laser hole drilling in the aerospace industry is the laser-drilling of cooling passages in cast superalloy turbine blades and vanes. This application has been in production for several years, and millions of laser-drilled holes have seen service in aircraft gas turbine parts.

Another example of current laser utilization is the formation of a continuous weld in a small impingement tube for use in a gas turbine engine. A 400 watt cw neodymium-YAG laser is used for this purpose. In this application the ability to precisely control the level and location of energy input leads to significant processes advantages, i.e., a clean, strong joint with minimal thermal distortion is created by a fully automated tape-controlled process.

## General Commercial

The largest production welding application using high power lasers is a lead welding process applied to large lead-acid storage batteries. The weld pictured in the inset of Fig. 31 indicates that smooth welds are produced at speeds of 8.47 cm/sec at a power of 2 kW. This high welding speed is unmatched by other processes for the welding of lead.

The system is schematically illustrated in the main body of Fig. 31. The application requires precisely controlled energy in order to produce a partial penetration weld, which must exceed 90% of the material thickness, but never fully penetrate the lug. Close control of energy input is thus required. The high welding speed and minimal specific weld energy promotes rapid cooling, so that the lead cannot slump out of the weld zone during processing.

LASER PROCESSING OF MATERIALS 73

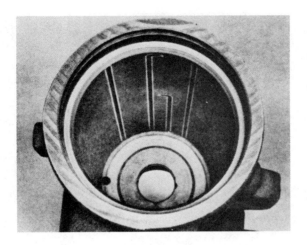

Fig. 30  Laser hardened wear stripes in bore of power steering pump housing.

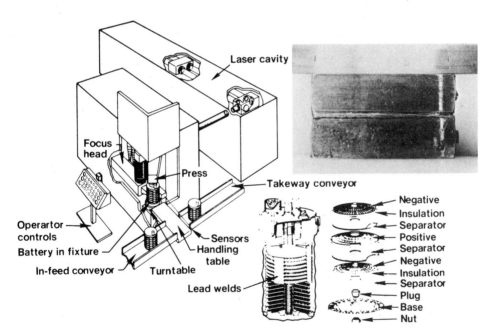

Fig. 31  3 kW Battery welding system and structure of butt weld in lead.

Fig. A  Stream of $Al_2O_3$ particles being dropped into a horizontal 6 kilowatt laser beam. Those particles which drop into the center of the beam are melted and propelled downstream.

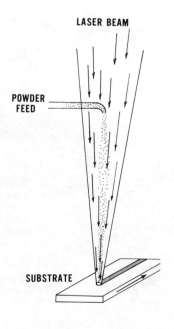

Fig. B  Schematic diagram of laser spraying system, showing powder being injected into laser beam via a water-cooled nozzle and deposited onto a moving substrate.

**LASER SPRAYING**

## CONCLUSIONS

Important applications of lasers for the processing of materials have been identified, and lasers are in use in production environments. The availability of laser systems with improved reliability will further enhance their industrial applicability.

The current developments in laser materials processing research are perhaps of greater significance than the current and near-term applications. The unique capabilities of this controllable, out-of-vacuum energy source will stimulate further research which in turn will solidify the commercial importance of the laser as a materials processing tool.

## APPENDIX

During the discussion period of the conference itself, B. B. Rath gave a brief report on a new process, termed 'laser spraying,' which is being developed by R. J. Schaefer and J. D. Ayers at the Naval Research Laboratory. For the sake of completeness, a summary of this work is given below.

Powder particles are injected by a CW laser beam of about five kilowatts. The particles are both heated and propelled by the laser beam. For example, aluminum oxide particles approximately 50 μm in diameter are rapidly melted and driven at velocities of a few meters per second through air. Fig. A. is a photograph showing $Al_2O_3$ particles dropped into a horizontal laser beam. The particles upon laser exposure are propelled in the direction of the beam. Particles which impinge upon aluminum or copper substrates have a clear glassy appearance indicative of rapid solidification. A schematic diagram of the laser spraying process is shown in Fig. B.

Corrosion and wear resistant coatings are sought by selection of appropriate powder chemistry. In air, the process is limited to materials such as oxides which are not degraded by high temperatures. For other materials, the process must be carried out in an environmental chamber with inert gas or vacuum. By adjusting the focal point of the laser beam, the heating or melting of the substrate can be varied, thus controlling the extent of alloying between powder and substrate materials.

## REFERENCES

1. Charschan, S. S., Editor: Lasers in Industry. Van Nostrand Reinhold Co., New York, 1972.

2. Ready, J. F.: Effects of High-Power Laser Radiation. Academic Press, New York, 1971.

3. Sullivan, A. B. J. and P. T. Houldcroft: Gas-Jet Laser Cutting, British Welding Journal, Aug. 1967.

4. Wick, D. W.: Lasercutting - Current Applications, Future Potentials, SME Paper MR74-965, 1974.

5. Belforte, A.: $CO_2$ Laser Cuts Metals - and Costs. Electro-Optical Systems Design, May 1975.

6. Alwang, W. G., L. A. Cavanaugh, and E. Sammartino: Continuous Butt Welding Using a Carbon-Dioxide Laser. Welding Research Supplement, Mar. 1969, p. 110.

7. Brown, C. O. and C. M. Banas: Deep-Penetration Laser Welding, Paper presented at the AWS 52nd Annual Meeting, San Francisco, CA, April 26-29, 1971.

8. Locke, E. V., E. D. Hoag and R. A. Hella: Deep-Penetration Welding with High-Power $CO_2$ Lasers. IEEE Journal of Quantum Electronics, Vol. QE-8 No. 2, Feb. 1972, p. 132.

9. Banas, C. M.: Laser Welding Developments. Proceedings of the CEGB International Conference on Welding Research Related to Power Plant, Southhampton, England, Sept. 17-21, 1972.

10. Adams, M. J.: Report #3335/3/73, British Welding Institute, Cambridge, England, 1973.

11. Baardsen, E. L., D. J. Schmatz and R. E. Bisaro: Welding Journal, April 1973.

12. Breinan, E. M. and C. M. Banas: Proc. 50th Anniversary Meeting of the Japan Welding Society, Osaka, Japan, 1975, p. 25, 137.

13. Yessik, M. and D. J. Schmatz: Laser Processing at Ford. Metal Progress, May 1975, p. 61.

14. Locke, E. V. and R. A. Hella: IEEE J. Quantum Electronics, QE-10, 179 (1974).

15. Seaman, F. D. and D. S. Gnanamuthu: Using the Industrial Laser to Surface Harden and Alloy, Metal Progress, Aug. 1975, p. 67.

16. Breinan, E. M., L. E. Greenwald, C. M. Banas, J. W. Davis, and J. P. Carstens: Proc. ASM Conference on Laser Surface Treatment for Automotive Applications, Detroit, Michigan, February 1976.

17. Miller, J. E. and J. A. Wineman: Metal Progress 110, 38 (1977).

18. Fairand, B. P., B. A. Wilcox and W. J. Gallagher, D. N. Williams, J. Appl. Phys. 43, (1973), p. 3893-3895.

19. Fairand, B. P., A. H. Clauer, R. G. Jung, B. A. Wilcox: Appl. Phys. Letters 25, (1974), p. 431-433.

20. Breinan, E. M., B. H. Kear, C. M. Banas and L. E. Greenwald, Proc. 3rd Int. Symp. on Superalloys, Seven Springs, Claitor's Publishing Div. Baton Rouge, 1976, p. 435-450.

21. Breinan, E. M., B. H. Kear and C. M. Banas: Physics Today, No. 1, 1976, p. 44-50.

22. Anthony, T. R. and H. E. Cline: J. Appl. Phys. 49, No. 3 (1978), pp. 1248-1255.

23. Khaibullin, I. B., E. I. Shtyrkov, M. M. Zaripov, M. F. Galyautdinov, and G. G. Zakirov, Sov. Phys. Semicond., 11, 190 (1977).

24. Brown, W. L., et al., Proc. of the Reston Conference on Rapid Solidification Processing, Claitor's Publishing Division, Baton Rouge, p. 127, (1977).

25. Leamy, H. J., G. A. Rozgonyi, T. T. Sheng and G. K. Celler: Appl. Phys. Letters 32, No. 9, p. 535, (1978).

26. Young, R. T., et al.: Appl. Phys. Letters, 32, #3, p. 139, (1978).

27. Young, R. T. and J. Narayan: Appl. Phys. Letters, June (1978).

28. Narayan, J., R. T. Young, R. F. Wood and W. H. Christie: Appl. Phys. Letters, Aug. (1978).

29. Copley, S. M., M. Bass and E. Garnuic: Laser Assisted Hot Spot Machining, Contract reports for N00014-77-C-0478.

30. Christensen, C. P., and K. M. Lakin: Appl. Phys. Letters, 32, #4, p. 254, (1978).

31. Tuli, M., P. R. Strutt, H. Nowotny, and B. H. Kear, Proc. of Conference on Rapid Solidification Processing, Reston, VA (1977).

32. Snow, D. (private communication).

33. Clauer, A. H., B. P. Fairand and B. A. Wilcox: Met. Trans. A, 8A, 1871, (1977).

34. Clauer, A. H., B. P. Fairand and B. A. Wilcox: Met. Trans. A, 8A, 119, (1977).

ELECTROHYDRODYNAMIC TECHNIQUES IN METALS PROCESSING

Julius Perel[*], John F. Mahoney[*], Bernard E. Kalensher[*] and Robert Mehrabian[**]

[*]Phrasor Technology, Duarte, California 91010
[**] University of Illinois at Urbana-Champaign
     Urbana, Illinois 61801

## 1. INTRODUCTION

The aim of this work is to develop a table top rapid quenching metal powder generator as a fundamental tool for the laboratory scientist permitting the development of new alloy compositions. The table top generator utilizes electrohydrodynamic (EHD) processes to generate the powder with a goal to produce more than 20 grams/day at cooling rates exceeding $10^5$K/sec for use as a laboratory research tool. The advantages of the process lie in it being a continuous process, having no vital moving parts, and can include radiation, convection and splat cooling techniques. The initial accomplishment was a demonstration of the feasibility of utilizing EHD, in the production of a wide range of powder sizes and corresponding cooling rates.

Varying EHD process parameters permits the control of particle production to obtain sizes much smaller than that achieved in conventional atomization processes. The process lends itself to the rapid production of a variety of alloy compositions required for the fundamental studies relating cooling rates to structure, segregation, etc. This could permit the development of new alloy compositions that would exploit the advantages of rapid solidification processing.

The EHD technique lends itself to a variety of controlled undercooling experiments coupled with rapid solidification. Extremely fine atomized droplets (micron size and smaller) have been routinely produced in the EHD apparatus. Therefore, it should be possible to carry out classical undercooling experiments (where droplet size is small enough to assure the absence of impurity particles responsible for heterogeneous nucleation in a large fraction of the droplets) coupled with rapid solidification. Very fine particles generated by this apparatus has been directly examined by TEM techniques.

## 2. EXPERIMENTAL APPROACH

### 2.1 Review of Electrohydrodynamics

The electrohydrodynamic (EHD) method for droplet generation involves the use of very intense electric fields which are attained by exposing a conductive liquid to electrodes having high curvature geometry while using moderate voltages of 3 to 20 kilovolts. The interaction of the electrostatic stresses with surface tension forces results in a highly dynamic process at the charged liquid surface. When the outward force exerted on the liquid miniscus exceeds the surface tension forces, the surface forms liquid jets or spikes required to produce the very high electric fields needed for charged droplet generation. After generation the droplets are then accelerated by the applied voltage.

Controllable variables include the acceleration voltage and the liquid metal feed rate. The properties of the alloy composition are also part of the variables affecting the results. The droplet trajectory and impact energy are controllable by means of electric and magnetic fields. In general, the droplet size can be increased by decreasing the electric field (voltage) and/or by increasing the flow rate.

The Table Top Model will be preceded by Laboratory Feasibility Model which is being used to demonstrate feasibility of the process and to determine the critical operational parameters in addition to initial materials investigations.

Materials processing was accomplished in the chamber of the Feasibility Model which consists of a cylindrical vacuum vessel which is 1 meter long and $\approx 0.3$ meter in diameter. The chamber can be evacuated to an operational level ($10^{-5}$torr) in a short time (1200 to 1800 seconds) using one 0.1525m (6 inch) diffusion pump. In its present arrangement, the processing vessel is configured so that the droplet beam is ejected horizontally. For symmetry, ease of collection, and other reasons, it appears to be more advantageous to orient the processing chamber in a vertical position.

## 2.2 Source Development

The source used for generation of microdroplets in the EHD apparatus is by far the most important component of the system. In addition to being versatile and inexpensive, it should fulfill the following criteria:

a. Show little or no reactivity with the molten alloy stream

b. Operate at a predetermined temperature for a given alloy system

c. Permit accurate control of flow through the nozzle

During the course of this program, four microdroplet sources were designed, fabricated, and tested. These were constructed from the following materials:

a. Stainless steel with a tungsten nozzle

b. Stainless steel - ceramic hybrid

c. Quartz with a ceramic nozzle

d. Ceramic

The stainless steel sources were the first source types fabricated to test the overall EHD concept in generation of microdroplets with a low temperature model Sn-15% Pb alloy. Fine metal powders (droplet sizes less than 1 µm) were successfully generated with these sources by electrohydrodynamic emission.

To pretest concepts leading toward an all-ceramic, high temperature, microdroplet source design, a hybrid source system consisting of stainless steel and $Al_2O_3$ feed and nozzle components was fabricated and successfully tested. $Al_2O_3$ cement was used as bonding agent between the different components of the source. Tests were performed with superheated Al-4.5% Cu alloy. Feed tube and nozzle dissolution problems noted earlier with the completely stainless steel source were absent in the hybrid source.

An inexpensive source made from quartz was a one-piece design molded from tubing. The crucible, feed tube, nozzle, and sealing flange were all glass blown into an integral structure. The overall simplicity and low cost of fabrication was offset by its fragility and perhaps limited versatility in containing a wise selection of alloys. However, the powder producing tests were successful and this source material could be used in special cases.

Finally, an all-ceramic source was developed and successfully tested. The basic design is flexible enough to accommodate construction of the source from a variety of ceramic materials, including mullite, alumina, and beryllia. Nozzles with desired orifice diameters of 25 to ≈ 175 μm of these materials are available. The initial all-ceramic source for the Feasibility Model was of mullite. Figure 1 shows photographs of this source which was successfully used to generate droplets of the Aℓ-4.5% Cu alloy.

## 3. EXPERIMENTAL RESULTS

All four sources were tested in the Feasibility Model EHD apparatus using a low temperature model alloy, Sn-15% Pb, and an aluminum alloy, Aℓ-4.5% Cu. The Sn-15% Pb alloy was extensively used in bench-type tests to verify the electrohydrodynamic process for generating fine liquid droplets, as well as to develop the necessary process variables for operation of the apparatus. The Aℓ-4.5% Cu alloy was also extensively utilized in preliminary

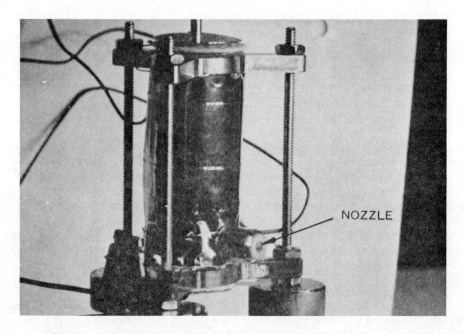

Figure 1. Photograph of the ceramic (mullite) source in the Feasibility Model.

# ELECTROHYDRODYNAMIC TECHNIQUES

experiments to:

a. Demonstrate that the EHD apparatus can be successfully used to generate a large range of fine and coarse atomized droplets (from less than 0.1 μm to over 100 μm).

b. Carry out a preliminary analysis of the range of structures, hence cooling rates, achievable in the apparatus.

c. Develop techniques for collection of the fine atomized powders.

d. Demonstrate the capability to produce rapidly solidified splats on various substrates.

e. Investigate the possibility of producing thin, adherent, rapidly solidified coatings on substrates.

f. Develop specific specimen preparation procedures for examination of the resulting structures by electron microscope techniques.

## 3.1 Atomization Studies

It was successfully demonstrated that the EHD apparatus can produce a large range of atomized droplets with each of the four source types. In these preliminary studies, heat flow during flight of the droplets was by radiative cooling only. Therefore, depending on the size of the droplets, their velocity, and the location of the collector, the droplets were partially or completely solidified in flight or arrived at the target as liquid.

Figure 2 shows Secondary Electron Images (SEI) of Aℓ-4.5% Cu powders collected on a vinyl cellulose substrate located in the flight path. Essentially fine atomized powders in the size range of less than 0.1 to 1.0 μm are shown. A soft target material, i.e., vinyl cellulose, was successfully used to collect the very fine particles generated by EHD. Furthermore, the spheroidal geometry of most of the powders indicates that these were probably solidified in flight.

Figure 3 shows a secondary electron image of the coarse powders, 100 to 150 μm size range, removed from the collector. The partially flattened particles showing rippled surfaces probably contained low volume fractions of solid prior to impact. The

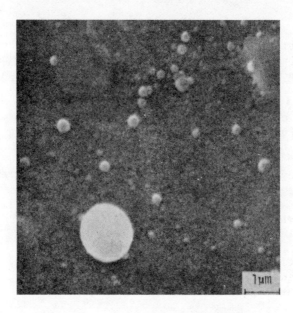

Figure 2. Secondary Electron Images of fine atomized powders of Aℓ-4.5% Cu alloy produced by the EHD technique. These particles were rapidly solidified in flight and collected on substrate material of vinyl cellulose.

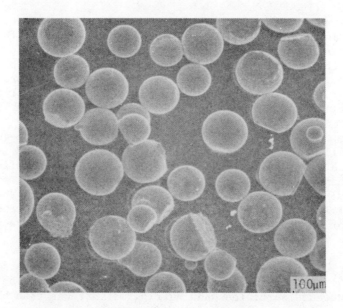

Figure 3. Scanning Electron Images of coarse atomized, 100 to 150 μm, powders of Aℓ-4.5% Cu alloy.

spherical particles, with some minor surface irregularities due to impact, probably also solidified after contact with the collector. Dendritic structure on the surface was clearly distinguished on these particles.

Finally, initial TEM Studies indicate that the very fine particles can be directly studied by this technique without any thinning of the powders. Preliminary TEM studies show that a substantial portion of the 1 µm, or less, diameter particles are single crystalline as illustrated in Figure 4.

### 3.2 Splat Cooling Studies

A second important capability of the EHD apparatus is its ability to produce controlled thickness, rapidly solidified splats of alloys on a variety of substrate materials. Splats of various thicknesses can thus be obtained, depending on the flow rate of the alloy, the fractional area of the jet covered by the target,

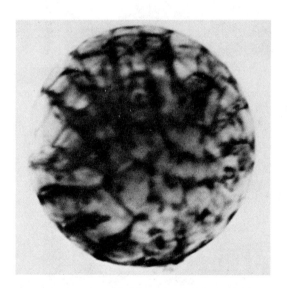

Figure 4. Transmission Electron Micrograph of a small powder (diameter < 1µm) of Aℓ-4.5% Cu alloy produced by the EHD technique. Magnification: 75,000X. Selected Area diffraction pattern taken from the powder indicates that it is a single crystal. The structure noted appears to be cellular - all the cells are part of a single grain.

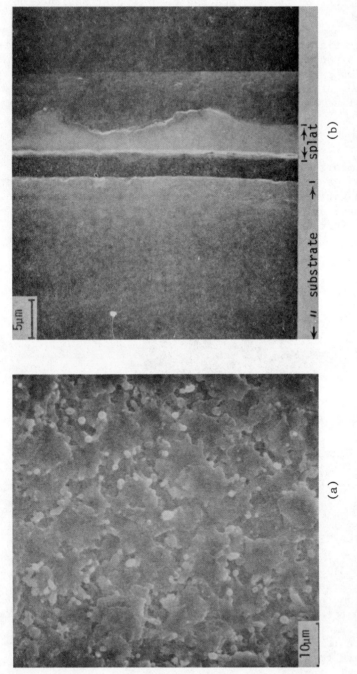

Figure 5. Secondary Electron Images of a splat cooled specimen of Aℓ-4.5% Cu alloy produced in the EHD apparatus on an aluminum substrate; (a) shows a view of the splat surface at 1000X; (b) shows a cross-sectional view of the splat and the substrate at 3000X. (Reduced 10% for reproduction.)

and its residence time. On impact, the high velocity liquid droplets flatten out on the substrate. Successive impact of liquid droplets results in a rapidly solidified, splat cooled foil of the alloy. In the preliminary experiments performed, three different substrate materials were employed for splat cooling of the Aℓ-4.5% CU alloy. These were thin, ≃ 100 μm to 500 μm, foils of vinyl cellulose, aluminum, and copper.

Figure 5 shows both top and edge views of a splat cooled specimen produced on an aluminum, 100 μm thick foil, substrate. It is evident that successive droplets were welded together in a manner similar to classical splat cooled specimens - no layering effects were observed. The relatively uniform (2 to 3 μm thick), featureless, cross-sectional view of the splat in Figure 5(b) shows that the EHD technique is an ideal method for preparation of rapidly solidified splat cooled specimens for subsequent fundamental studies via electron microscopy techniques. Absence of a distinguishable microstructure at magnifications of up to 10,000X in the splats produced to date, indicates that either cooling rates were high enough to produce a complete solid solution of the alloy, or that the substructure is so fine that it could not be observed at these magnifications.

A further aim of this study is to investigate the possibility of deposition of adherent thin films via the EHD technique. Figure 6 shows some evidence of welding between a splatted foil of Aℓ-4.5% Cu alloy and an aluminum substrate.

## 4. CONCLUSIONS

This work has demonstrated the concept feasibility of a Table Top Amorphous/Microcrystalline Powder Generator using electrohydrodynamic methods. Accomplishments attained from these investigations are briefly summarized as follows:

a. A ceramic particle source was developed and operated at 1000°C and laboratory tested to 1300°C.

b. Submicron particles (less than 0.1 micron diameter) were generated from an Aℓ-4.5% Cu alloy.

c. Preliminary TEM analysis indicated that a large quantity of the submicron Aℓ-4.5% Cu particles were single crystalline.

d. Cooling rates of $10^5$ to $10^6$ °C/second were determined for particles solidified in flight before collection. Splat cooling undoubtedly has higher cooling rates.

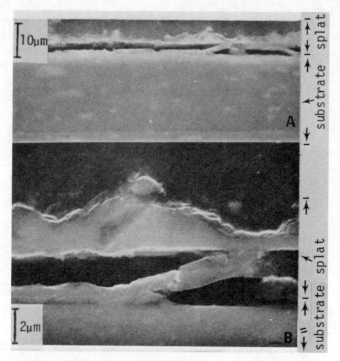

Figure 6. Cross-sectional view of an Aℓ-4.5% Cu alloy splat cooled specimen deposited on an aluminum substrate showing evidence of welding between the two. Top and bottom Secondary Electron Images are at 1000X and 5000X, respectively. (Reduced 10% for reproduction.)

e. Relatively uniform coatings (several micron thick) were deposited on various substrates using the Aℓ-4.5% Cu alloy.

f. Photomicrographic examination of a ceramic (alumina) nozzle used in Aℓ-4.5% Cu tests shows no corrosive interaction.

The selection of the electrohydrodynamics approach for the production of amorphous substances, metastable phases, and materials showing increased solid solubility of solute elements is gaining in importance as an advanced materials processing technique. This work is complemented by the efforts of other investigators who recently demonstrated techniques for forming solid bars from metallic glass powder without destroying their useful properties. The advent of fabrication methods that will provide useful shapes to industry and research facilities, starting with powders, greatly enhances the possibility of the practical utilization of the EHD powder generation method.

## ACKNOWLEDGEMENT

This research is sponsored by the Defense Advanced Research Projects Agency and monitored by the Office of Naval Research under Contract #N00014-77-C-0373. Electron Microscope characterization of the specimens by C. G. Levi is gratefully acknowledged.

FUNDAMENTALS OF PARTICULATE METALLURGY

Alan Lawley
Department of Materials Engineering
Drexel University
Philadelphia, Pa. 19104

ABSTRACT

High performance powder metallurgy reflects significant advances in the science and technology of powder production and subsequent powder consolidation. Powder processing methods involving the atomization of a stream of liquid metal are reviewed with particular reference to relationships between processing parameters and powder properties. These methods include gas and water atomization, several forms of centrifugal atomization and recent innovative approaches at the prototype or laboratory-stage of development. Where available, models and mechanisms of the atomization processes are detailed. To conclude, the role of cooling rate in powder processing, and implications re the characteristics of rapidly solidified powders are examined.

INTRODUCTION

The Sagamore Army Materials Conference held in 1971 focussed on powder metallurgy for high-performance applications. Since that time, a better understanding of the various technologies involved has been achieved. Also, the spectrum of materials amenable to powder processing to full density has increased significantly. In the area of powder production most of the interest has centered on liquid metal atomization (1). More refined empirical relationships between processing parameter and powder properties have been documented and the general level of powder quality, in terms of size, shape, chemistry and purity, has improved (1). Innovative techniques of atomization now exist which may become challengers to the conventional processes of gas or water atomization (2,3). Perhaps the most significant progress has been made in the area of atomization mechanisms. Understanding of the fundamentals of the atomization process provides a basis for optimal design vis a vis powder quality and yield.

Stimulus for further development and understanding of high performance powder metallurgy resides in factors such as:

- materials conservation
- compositional flexibility
- fine grain size
- isotropy
- energy conservation
- homogeneity (structure, composition)

The initial step involves the preparation of a suitable powder. While there are several broad approaches of commercial importance (e.g. chemical, electrolytic, comminution, atomization) it is in the area of liquid metal atomization that most research has been directed over the last decade – with the objective of increasing yields, lowering particle size, tightening chemistry, and reducing impurities such as oxide inclusions. The thrust toward atomization reflects the need to develop viable powder processing routes for materials such as alloy steels, titanium and the superalloys.

In this paper, commercial and more recent prototype-scale methods of liquid metal atomization are reviewed. Consideration is given to processing variables, resulting powder properties and operative mechanisms of atomization. To conclude, the effect of cooling rate on particle structure is examined covering the range from gas atomization ($\sim 10^2 °C/s$) to rapid solidification processing ($\gtrsim 10^5 °C/s$).

## ATOMIZATION METHODS

### Gas Atomization

The principle of gas atomization is simple – a continuous stream of liquid metal is broken down into droplets by means of a subsonic or supersonic gas stream or jet. Atomization is due to the kinetic energy transfer from the atomizing medium to the metal. In practice, nitrogen, argon or air are used; mixtures of these gases and helium are also effective. The number and geometry of gas-metal configurations are unlimited; typical arrangements involving multiple jets or an annular ring are illustrated schematically in Figure 1. More detail concerning jet configurations and design of atomizing units is available in the open literature (4-9).

There are several interrelated processing and material variables involved in gas atomization, Table I. In consequence

# FUNDAMENTALS OF PARTICULATE METALLURGY

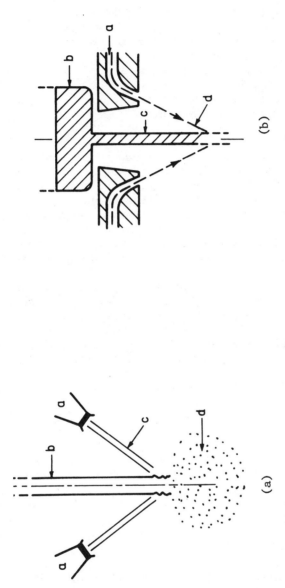

Figure 1. Representative gas-atomization configurations: (a) Two-jet configuration; a = jets, b = liquid metal stream, c = gas stream, d = atomized powder. (b) Annular-ring configuration; a = ring orifice, b = liquid metal reservoir, c = liquid metal stream, d = gas stream.

## Table I

### Variables in Gas Atomization

Gas jet distance and pressure
Gas velocity and mass flow rate
Metal velocity and mass flow rate
Nozzle geometry
Angle of impingement
Superheat
Metal surface tension
Metal melting range

numerous empirical relationships have been documented or proposed for the prediction of particle size and shape and the distributions of size and shape (1,9). It is always found that as gas pressure increases, as gas mass flow rate increases and/or as the jet to metal stream distance decreases, the average particle diameter decreases (8-12). The power law relating particle size to gas pressure is shown in Figure 2. Typical operating ranges of some of the above variables in gas atomization are summarized in Table II.

Gas-atomized powders are typically spherical and have smooth surfaces. Particle cooling rate is $10^2$°C/s. A representative size distribution for superalloy powders is shown in Figure 3, curve a. The mean particle diameter is ∼150μm; yields up to ∼40% are achieved at -325 mesh size. Alloy steels, titanium and superalloy compositions are amenable to inert gas atomization (13-16). Overall, the process has a low energy efficiency and is expensive if gases other than nitrogen have to be used.

### Water Atomization

Water atomization is similar to gas atomization in that it is a two-fluid process; geometric arrangements of the water stream relative to the liquid metal stream are similar to those used in gas atomization, Figure 1. Process parameters correspond to those

## Table II

### Representative Operating Conditions in Gas Atomization

| | |
|---|---|
| Gas Pressure | $14 \times 10^5 - 42 \times 10^5$ Pa |
| Gas Velocity | 50-150 m/s |
| Superheat | 100-200°C |
| Angle of Impingement | 15-90° |

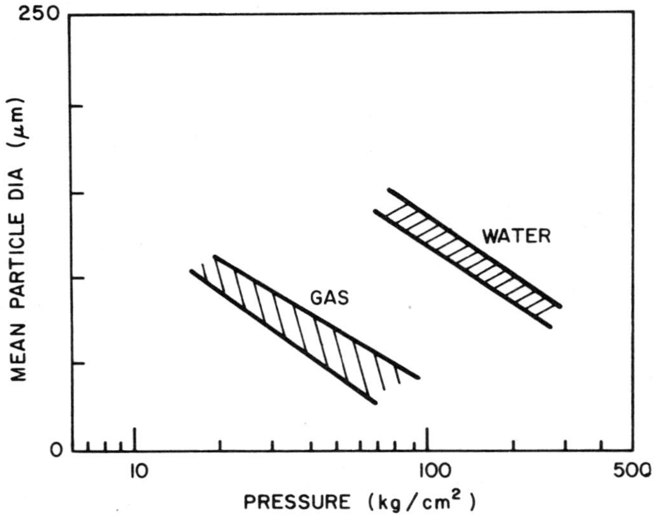

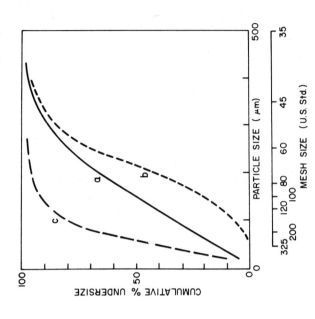

Figure 2. The dependence of particle size on fluid (gas or water) pressure. 1 kg/cm$^2$ = 97.9 x 10$^3$Pa. (11)

Figure 3. Representative particle size distributions for powders prepared by: a) Inert gas atomization, b) The rotating electrode process, c) Vacuum atomization. (13)

## Table III

### Representative Operating Conditions in Water Atomization

| | |
|---|---|
| Water Pressure | $35 \times 10^5 - 210 \times 10^5$ Pa |
| Water Velocity | 40 - 150 m/s |
| Superheat | 100-250°C |
| Angle of Impingement | $\leq 30°$ |

listed in Table I for gas atomization. Representative operating conditions in water atomization are listed in Table III. Water pressure is usually higher than the pressures adopted in gas atomization. Generally, the angle of impingement is smaller for water than for gas atomization. Specific details on water atomization units are available in the literature (7,9,10,17-21).

Large quantities of energy are required to supply the water at high pressure and the overall energy efficiency is very low, $\sim 4\%$. The amount of water used per unit weight of powder produced is extremely sensitive to the geometric arrangement of the water jet relative to the liquid metal stream. The process is significantly cheaper than gas atomization unless air can be used as the gas medium.

Water atomized powders are generally irregular in shape with rough oxidized surfaces. Low and high-alloy steels, including stainless, are amenable to water atomization. Excess oxide on the powder particle surfaces can be subsequently removed or diminished in amount by hydrogen reduction or various forms of acid leaching.

It has been established empirically that small particles are favored by high water pressure, high water velocity, a short jet to metal distance and a low metal flow rate. The dependence of particle size on water pressure is shown in Figure 2. Superheating of the metal also decreases particle diameter, primarily through the associated decrease in the surface tension of the metal at the actual atomizing temperature. In terms of particle shape, some control can be exerted in water atomization. Powder particles become less irregular and approach sphericity as the degree of superheat is increased and as the angle of impingement of the water jet is increased. Alloys with a narrow melting range (i.e., a small temperature interval between liquidus and solidus) result in some degree of sphericity. Powder yields increase as the water pressure increases. Yields $\sim 60\%$ of -35 mesh powder can be achieved at a water pressure of $60 \times 10^5$ Pa. Particle cooling rates ($\sim 10^3$°C/s) are an order of magnitude higher than those achieved in gas atomization.

Steam atomization is a variant of gas and water atomization. The process is relatively cheap and can be used in the production of several classes of alloys: carbon and low alloy steels, stainless steels, nickel and cobalt-base superalloys. Subsequent treatment of steam atomized powders is usually necessary to lower surface oxide content. This may take the form of milling or chemical leaching. The virtues of low-pressure steam atomization could result in expanded use of this technique on a commercial scale.

## Centrifugal Atomization

In centrifugal atomization, molten metal is ejected in the form of droplets from a rapidly spinning crucible plate or disc. Particles either cool in the environment of the collection chamber or can be gas-quenched as they leave the rotating vehicle. Many configurations have been examined (22-26). The arrangement illustrated schematically in Figure 4 is suitable for titanium alloy powders. Erosion, dissolution, oxidation and creep of the spinner pose severe materials problems; the advent of improved ceramics has been a key factor in the development of this process. Energy requirements are relatively low.

Particle shape can be varied from spherical to flake. Spherical powders produced by centrifugal atomization are relatively coarse with average particle diameters $\gtrsim 200 \mu m$.

Other forms of centrifugal atomization include the rotating electrode process (13,27), the Durarc® process (28) and the several techniques termed melt spinning and melt extraction. Since spinning and extraction from the melt produce primarily filaments or ribbons, these techniques will not be considered further. Maringer et al. (29,30) have recently described melt extraction and melt spinning. Cooling rates in excess of $10^6 °C/s$ are reported. Amorphous and/or microcrystalline materials in filament or ribbon form offering unique magnetic properties are now produced on a commercial scale by melt extraction.

The rotating electrode process (13,27) illustrated in Figure 5 is in commercial operation though production rates are low. Material in the form of a rod electrode is rotated rapidly while it is melted at one end by an electric arc. Molten metal spins off the bar and solidifies before hitting the walls of the inert gas-filled chamber. The process has been developed primarily for atomization of titanium alloys and for superalloys. Tungsten contamination from the stationary electrode can be avoided by replacement with a titanium cathode; both cathode and anode rotate in the most recent version.

Rotating electrode powders are usually spherical, of high surface quality but have a large average particle diameter $\gtrsim 200 \mu m$. Size range is $\sim 50$-$400 \mu m$, Figure 3 (curve b). Typically, yields run

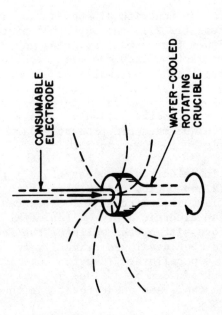

Figure 4. Schematic representation of a centrifugal atomization configuration.

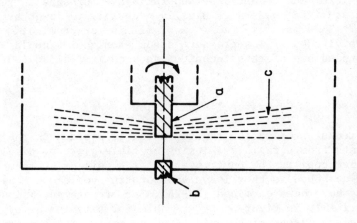

Figure 5. A schematic to illustrate the rotating electrode process. a and b are electrodes, c = powder.

90% for -35 mesh powder. Cooling rate is low and comparable to that achieved in gas atomization, i.e. $\sim 10^2 °C/s$.

### Other Atomization Techniques

A variety of atomization techniques has been reported in the literature which do not fit the categories of gas, water, or centrifugal atomization. These range from laboratory scale curiosities to prototype or semi-production units. Three of the most advanced/promising methods will be considered.

In roller atomization (31), a stream of molten metal is fed between rapidly rotating rolls (up to 200 rev/s). A schematic is given in Figure 6. The roll gap is $\sim 50\mu m$. Heat transfer to the rolls is minimized by a suitable insulating coating so that the atomization event actually takes place by break up of the molten sheet of metal below the roll gap. Singer (31) has examined process variables for copper, tin and lead. A median particle size of 220μm has been obtained for tin powders with a yield $\sim 70\%$. Median particle size is inversely proportional to roll speed. A wide spectrum of forms can be atomized: flake, acicular, irregular, spheroidal. The method is attractive in terms of energy consumption and should be amenable to scale-up. Cooling rates have not been determined.

Ultrasonic atomization appears attractive for the production of small diameter (<50μm) particles with a narrow size distribution (3) at high quench rates ($\sim 10^5 °C/s$). Hartman shock wave tubes are used to deliver high velocity pulses of gas (up to Mach 2) at frequencies in the range 60,000-80,000 cps. There is no contact between the ultrasonic 'die' and the liquid metal stream and the configuration is similar to that in gas atomization. Grant (3) has obtained yields in the 80-90% range for -325 mesh powder of stainless steel. Ultrasonic atomization has been used on a commercial scale for low melting point alloys and more recently interest has centered on superalloys. Like roller atomization, there do not appear to be any major deterents to scale-up.

Vacuum or soluble gas atomization (13,32) is a commercial batch process based on the principle that when a molten metal supersaturated with gas under pressure is suddenly exposed to vacuum, the gas expands, comes out of solution, and causes the liquid metal to be atomized. Alloy powders based on Ni, Cu, Co, Fe and Al have been vacuum atomized with hydrogen as the gas. The powders are spherical, clean and are of a high-purity relative to other powder processing methods. A schematic is given in Figure 7 and a typical particle size distirubtion for a superalloy in Figure 3 (curve c).

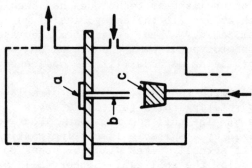

Figure 7. Schematic representation of vacuum atomization: a = trap door, b = transfer tube, c = molten metal.

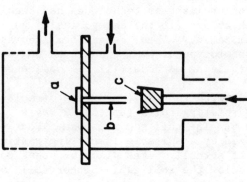

Figure 6. Schematic representation of roller atomization: a = liquid metal stream, b = heat source, c = rolls, d = metal powder.

# FUNDAMENTALS OF PARTICULATE METALLURGY

## ATOMIZATION MECHANISMS

Two-fluid atomization does not lend itself to simple modelling or analysis. However, several models have been advanced for explaining the sequence of events leading to liquid droplet formation from the metal stream. These pertain to gas, water, ultrasonic and roller atomization.

### Gas Atomization

The basis of the current understanding of the fundamentals of gas atomization must be credited primarily to See et al. (12,33,34). In a series of elegant experiments these workers utilized a variable area meter to map gas flow rate around the liquid metal stream and Pitot tubes to generate gas velocity profiles. The velocity measurements were complemented by Schlieren photography of the gas-metal stream interactions. It is shown convincingly that the overall process of gas atomization is a three-stage process, as illustrated in Figure 8. Primary break-up of the liquid stream occurs in stage I. In stage II the droplets are still molten so that secondary disintegration is possible. Finally in stage III, the particles solidify. It is important to note that the actual atomization event takes place above the point of focus of the gas stream, Figure 8. The reason for this resides in the vacuum created around the metal stream coupled with the stream acceleration which causes the liquid metal to expand into a hollow conical configuration of molten metal. The actual atomization event then occurs at the circumferential periphery of the cone.

What are the details of break up of the hollow cone formed during stage I? Here there is general agreement that a sinuous wave is initiated which increases rapidly in amplitude. Then the wave becomes detached from the bulk liquid, producing a ligament whose dimensions depend on the wavelength at disintegration. Various physical models are available with which to describe these events; stability criteria all involve the Weber number of the liquid metal. The model developed by Dombrowski and Johns (35) for the disintegration of a liquid sheet is illustrated in Figure 9. See, Runkle and King (34) analyzed their experimental data on the nitrogen atomization of lead in terms of this model of ligament formation and break up. From the theory and the known (experimental) values of mass mean particle diameter, a parameter termed the mean relative gas-wave velocity (U) was calculated and compared to the measured values of U parallel to the surface of the hollow cone at the point of atomization. Reasonable agreement existed. The several experimental studies of gas atomization (8,9,12,33,36) involving high-speed photography confirm that break-up of the liquid metal stream involves ligament formation. Subsequently, the ligaments break up into more spherical particles before solidification.

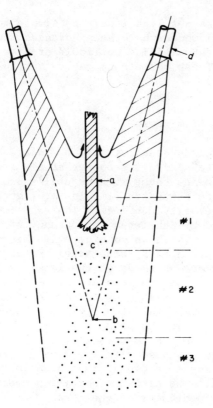

Figure 8. The three stages of liquid metal stream disintegration in gas atomization (12,33).

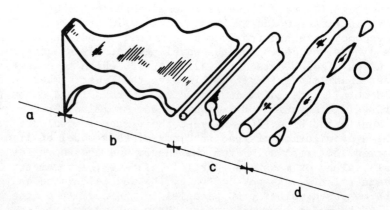

Figure 9. A model to show disintegration of a sheet of liquid: a = stable sheet, b = growth of waves in sheet, c = ligament formation, d = ligament breakdown (35).

# FUNDAMENTALS OF PARTICULATE METALLURGY

Bradley (37) has developed a rigorous mathematical model of gas atomization for the prediction of drop size at subsonic and sonic gas velocities; the model also takes into account the compressibility of the gas and the viscosity of the liquid. It is based on the physical model of ligament formation and break-up proposed by Castleman (38). From Bradley's analysis it is clear that compressibility of the gas stream is important - to the extent that drop size is dependent only sensibly on the Mach number of the gas flow and on the surface tension of the molten metal. Drop size decreases with increasing gas velocity but is linearly dependent on surface tension. The dimensional analysis leads to a single universal curve for the prediction of powder particle size as a function of gas velocity. This curve is given in Figure 10. M is the Mach number and L is a function of $\rho$, M and N where:

$$\rho = \rho_G/\rho_L; \quad M = U/U_S; \quad N = T/\rho_G U_S V \tag{1}$$

and
$\rho_G$ = gas density  $\quad$ U = gas velocity
$\rho_L$ = liquid density  $\quad$ $U_S$ = sonic gas velocity
T = surface tension of liquid  $\quad$ V = kinematic viscosity of liquid

To predict the diameter of the most frequently occuring particle for a particular gas velocity (in terms of Mach number), the corresponding value of L is obtained from Figure 10. Particle diameter is then given by:

$$d = \frac{2.95\ T}{L\rho_G U_S^2} \tag{2}$$

A rigorous test of Bradley's analysis awaits more extensive experimental data of the type developed by See et al. (12,33,34).

## Water Atomization

There have been only a few studies aimed at elucidating the atomization mechanism in water atomization. As a general observation, there is no evidence to suggest that ligaments occur as an intermediate step during particle formation.

Grandzol (21) has shown that a simple functional relationship exists between median particle diameter $d_m$ and water velocity $V_W$. Of importance here is the observation that the water breaks up into droplets prior to impacting the liquid metal stream; hence $V_W$ refers to water droplet velocity. The relationship is of the form:

$$d_m = K/V_W \tag{3}$$

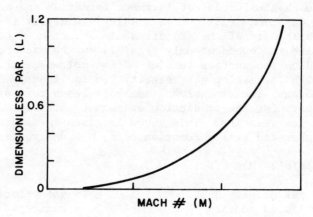

Figure 10. Bradley's universal plot for the determination of particle size (37).

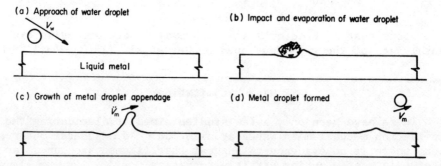

Figure 11. The 'scrape' mechanism of water atomization (21).

# FUNDAMENTALS OF PARTICULATE METALLURGY

Grandzol (21) and Grandzol and Tallmadge (39) also derived a relationship of this form by means of a simple two-liquid atomization model involving: momentum-exchange for determination of droplet size, empirical relations to relate water droplet size to water jet velocity, kinetic energy exchange for the ratio of water droplet velocity to metal particle velocity, and high-speed photography for the determination of the actual metal particle velocity. From their experimental observations, Grandzol and Tallmadge (40) also showed that it is the component of water velocity normal to the metal stream that is the dominant factor that controls metal particle size. Specifically:

$$d_m = \frac{K}{V_W \sin \alpha} \qquad (4)$$

where $\alpha$ is the angle between the metal-stream axis and the water jet axis (i.e. half the total impingement angle). This gives credence to a mechanism of droplet formation that involves impact as opposed to shear, as compared to gas atomization. A possible sequence of events consistent with the above model is illustrated schematically in Figure 11. This so-called "scrape" mechanism allows for almost instantaneous formation of the metal droplet. The momentum-interchange model is simple and does not require a one to one ratio between water droplets and resulting metal droplets.

Clearly, more work is needed to evaluate the generality of the Grandzol and Tallmadge model. The importance of water and steam atomization as a low-cost process for powder production on a commercial scale should serve to stimulate such research.

## Ultrasonic Atomization

In contrast to gas atomization, ultrasonic atomization appears to be a single step process (3). Because of the high velocity of the gas pulse, the liquid metal stream responds on impact as though it were a solid with low shear resistance. Preliminary calculations indicate a low overall efficiency, due primarily to energy losses by reflection of the gas pulse and through heat generation.

## Roller Atomization

Roller atomization is amenable to modelling and analysis with reference to break up of the liquid metal sheet as it moves downward past the roll gap (31). From high-speed photography, it is concluded that the actual atomization mechanism is analogous to that occuring in lubricant films; it is known as cavitation. Thus, the liquid metal actually cavitates a small distance downstream from the nip of the rolls and gives rise to perforations and streamers.

At high roller speeds there is evidence that the liquid metal flows around the points of cavitation and forms thin threads or slivers of liquid. These subsequently break up due to the propagation of liquid-jet instabilities. On-going research (41) is aimed at a quantitative description of the primary break up of the liquid sheet by cavitation and subsequent secondary break up and coalescence of liquid streamers and perforations. This involves establishing the correct boundary conditions for cavitation and solving the Reynolds equation taken from lubrication theory.

## RAPIDLY SOLIDIFIED POWDERS

Since the 1971 Sagamore conference on powder metallurgy increasing interest has developed in the area of rapidly solidified powders (RSP) - to the extent that a recent conference was devoted exclusively to the principles and technologies associated with rapid solidification processing (42). The central tenet is that rapid cooling rates offer unique possibilities in terms of the development and control of microstructure in metal powders. To put this in perspective, dendrite arm spacing, which is a measure of the scale of the liquid-to-solid segregation pattern, is listed as a function of cooling rate in Table IV. In turn, the cooling rate is controlled by the physical size of the atomized metal powder particle. Obvious benefits appear to exist in terms of a minimum of segregation within individual powder particles, inhibition of embrittling compounds or films (i.e. a greater tolerance for impurities) and ultrafine grain size.

Two conditions which must be met in order to rapidly solidify powders are small droplet size and high heat transfer at the metal droplet interface. To this end, Cox et al. (26,43) have developed a unique centrifugal atomization facility in which particles are cooled in a blast of helium as they leave the periphery of a rotating disc. In order to achieve cooling rates in excess of $10^5$°C/s, particles sizes must be below $\sim 60\mu m$ diameter and this requires rotational speeds in the range 250-600 rev/s. Some

Table IV

Dendrite Arm Spacing as a Function of Cooling Rate

| Cooling Rate (°C/s) | Dendrite Arm Spacing ($\mu m$) | Process |
|---|---|---|
| $10^{-2}$ | $5 \times 10^2$ | conventional ingot solidification |
| $10^2$ | $10^2$ | gas atomization |
| $10^3$ | 5 | water atomization |
| $10^5$ | 1 | substrate cooling |
| $10^7$ | 0.1 | splat cooling |

experimental data for IN100 are shown in Figure 12; cooling rate was determined from measured dendrite arm spacings in the powder particles.

At the present time RSP powders, primarily superalloy compositions, are being characterized with particular reference to structure (dendrite, microcrystalline). Interest also centers on possible changes in microstructure during consolidation to full density. There is a need to compare RSP and standard gas atomized powders consolidated under similar conditions; only in this way can the expected benefits of rapid cooling be truly appraised.

## SUMMARY

A broad spectrum of atomization techniques exist. Gas, water, steam and several forms of centrifugal atomization are commercially viable. Roller and ultrasonic atomization appear promising but require further development. Rapidly solidified powders ($>10^5$°C/s) are now available possessing unique microstructures; it remains to be seen whether or not the properties of consolidated RSP powders are superior to those derived from a conventional gas atomized powder. Significant progress has been made toward an understanding of the mechanism(s) involved in gas and water atomization; this is essential in order to achieve optimal design of each process.

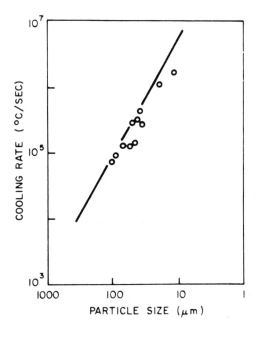

Figure 12. Experimentally determined cooling rates as a function of powder particle size (26).

## REFERENCES

1. A. Lawley, The Int. Journal of Powder Met. and Powder Tech., 13, #3, p. 169 (1977).
2. A. Lawley, "Preparation of Metal Powders" in Ann. Rev. Mater. Sci., Vol. 8, p. 49, Ann. Revs. Inc. (1978).
3. N. J. Grant, in Rapid Solidification Processing, Editors: R. Mehrabian, B. H. Kear and M. Cohen, Claitor's Pub. Div., p. 230 (1978).
4. H. Lubanska, J. Met. 22, 45-9 (1970).
5. C. F. Dixon, Can. Met. Quart. 12, 809 (1973).
6. E. Klar and W. M. Shafer, Powder Metallurgy for High Performance Applications, ed. J. J. Burke and V. Weiss, pp. 57-68, New York, Syracuse Univ. Press (1972).
7. P. U. Gummeson, Powder Met. 15, 67-94 (1972).
8. P. Rao, Shape and Other Properties of Gas Atomized Powders, Ph.D. Thesis, Drexel Univ., Philadelphia, Pa., 262pp (1973).
9. J. K. Beddow, The Production of Metal Powders, Heyden and Son, Ltd., Philadelphia, Pa. (1978).
10. S. Small and T. J. Bruce, Int. J. Powder Metall., 4, pp. 7-17 (1973).
11. P. Rao, R. G. Bowrey and J. A. Tallmadge, Chemeca '70, London, Butterworth, pp. 1-16 (1971).
12. G. H. Johnston and J. B. See, Formation of Liquid Metal Droplets, Paper presented at Symp. on Process Eng. of Pyrometall., Imperial College, London (1974).
13. L. P. Clark, Advanced Manufacturing Methods and their Economic Implications: Some Pilot Papers on Powder Metallurgy and Joining, AGARD Report #627, pp. 1-18 (1975).
14. P. B. Wallis, Powder Met. 8, 167-69 (1976).
15. J. S. Benjamin and J. M. Larson, J. Aircraft 14, pp. 613-23 (1977).
16. G. H. Gessinger and M. J. Bomford, Int. Met. Rev., 19, 51-76 (1974).
17. J. F. Watkinson, Powder Met. 1(2), 13-72 (1958).
18. O. Vorsa, Pikroky Praskove Metalurgie, 1, 47 (Brutcher translation #6572) (1963).
19. K. Tamura and T. Takeda, Trans. Nat. Res. Inst. Metals. 5 (5), 252-56 (1963).
20. P. U. Gummeson, See Ref. 6, pp. 27-55 (1972).
21. R. J. Grandzol, Water Atomization of 4620 Steel and Other Metals, Ph.D. Thesis, Drexel Univ., Philadelphia, Pa. 272pp (1973).
22. T. S. Daugherty, Powder Met. 11, 342-57 (1968).
23. H. Stephan, Advanced Fabrication Techniques in Powder Metallurgy and their Economic Implications, 42nd Meeting of the Structures and Materials Panel (AGARD), Ottawa, Canada, p. N77-15156 (1976).
24. P. W. Sutcliffe and P. H. Morton, See Ref. 23, p. N77-15157 (1976).
25. J. Decours, J. Devillard and G. Sainfort, See Ref. 23, p. N77-15154 (1976).

26. A. R. Cox, J. B. Moore and E. C. Van Reuth, Proc. 3rd Int. Conf. on Superalloys, Claitor's Pub. Div., pp. 45-53 (1976).
27. G. Friedman, See Ref. 23, p. N77-15155 (1976).
28. The Evaluation of Ti-6Al-6V-2Sn Pre-Alloyed Powder Processing, Westinghouse Electric Corporation Technical Report AFML-TR-76-65, 135pp (1975).
29. R. E. Maringer and C. E. Mobley, See Ref. 3, p. 208.
30. C. E. Mobley, R. E. Maringer and L. Dillinger, See Ref. 3, p. 222.
31. A. R. E. Singer and A. D. Roche, in Modern Developments in Powder Met., Editors: H. H. Hausner and P. W. Taubenblat, Vol. 9, p. 127, MPIF, Princeton, N. J. (1977).
32. J. M. Wentzell, See Ref. 23, p. N77-15165 (1976).
33. G. H. Johnston and J. B. See, Unpublished work, MIT (1974).
34. J. B. See, J. Runkle and T. B. King, Met. Trans. $\underline{4}$, pp. 2669-73 (1973).
35. N. Dombrowski and W. R. Johns, Chem. Eng. Sci. $\underline{18}$, pp. 203-14, (1963).
36. P. Karinthi, G. Raisson, Y. Honnorat and J. Morlet, See Ref. 31, pp. 115-25 (1977).
37. D. Bradley, Appl. Phys. $\underline{6}$, 1724-36, 2267-72 (1973).
38. R. A. Castleman, Jr., Nat. Bur. Stand. J. Res. $\underline{6}$, pp. 369-76 (1931).
39. R. J. Grandzol and J. A. Tallmadge, AIChE J. $\underline{19}$ (6) pp. 1149-58 (1973).
40. R. J. Grandzol and J. A. Tallmadge, Int. J. Powder Met. Powder Tech. $\underline{11}$ (1), pp. 103-14 (1975).
41. A. R. E. Singer, Private Communication (1977).
42. Rapid Solidification Processing, Editors: R. Mehrabian, B. H. Kear and M. Cohen, Claitor's Pub. Div. (1978).
43. P. R. Holliday, A. R. Cox and R. J. Patterson, See Ref. 3, p.246.

WELDING WITH HIGH POWER LASERS


E. M. Breinan and C. M. Banas

United Technologies Research Center
East Hartford, Connecticut   06108


INTRODUCTION & BACKGROUND

Lasers date back to 1960(1) which earmarked the initial operation of a ruby laser. For the following years, average laser output power remained low and, as a result, the development of materials processing applications progressed slowly. An important commercial application evolved in 1965(2) when the Western Electric Company adapted a pulsed system for production drilling of diamond wire drawing dies. In this application the processing time for drilling of a die was reduced from 24 hours for conventional techniques to less than one hour with lasers. A variety of commercial drilling applications have since been developed; one of the most successful being laser drilling of cooling passages in gas turbine parts (2,3). The success of these initial production applications hinged on the laser's ability to accurately deliver a precisely controlled amount of energy to a highly localized region of a workpiece.

Some of the first laser welding applications were spot welding of microelectronic components, as described in references 2 and 3. Seam welding was first accomplished by overlapping spot welds which were conveniently formed with available pulsed laser systems. The maximum attainable penetration in welds of this type was limited to approximately 0.060 in (1.5 mm) by the ten millisecond maximum pulse length characteristic of pulsed, solid-state laser systems. The penetration limit was imposed by attainment of vaporization temperature at the workpiece surface before appreciable energy could be conducted into the material. Maximum weld penetrations with pulsed lasers were obtained in materials having high thermal diffusivities.

The relatively shallow penetration limit coupled with the low welding rates which can be obtained with the overlapping spot technique has restricted the application of pulsed welding to small, thin-section assemblies such as microelectronic components which require precise control of welding energy.

The first solid-state lasers with the capability for continuous operation at substantial average power were yttrium aluminum garnet (YAG) and yttrium aluminate (YALO) (4). Power outputs to 1 kw have been achieved and have led to significant seam welding applications. The only unit currently suitable for heavy section industrial welding, however, is the carbon dioxide gas laser first developed by Patel (5) in 1964. Development of convective cooling and laser gas recirculation for the $CO_2$ laser have led to industrially suited units capable of continuous operation at maximum power output of the order of 20 kW (6-9).

Industrial laser applications have been facilitated by the development of high power cw units. Initial welding tests with cw carbon-dioxide lasers were conducted with lasers having output powers of less than 1000 watts. Maximum penetration capability of these lasers was found to be approximately 3mm (0.120 in) in stainless steel and welding speed was low (10-12). On the other hand, $CO_2$ lasers with several hundred watts output were found to be quite effective in cutting. Numerous applications were developed including slitting of steel rule dies, sheet metal pattern cutting, cutting of plastics and vinyls for the automotive industry, and computer controlled cutting of fabrics for the garment industry.

Development of multikilowatt cw laser systems opened the door to heavy section welding. The first results for laser welding with a mulikilowatt unit were reported in 1971 (13). The high power laser was shown to be capable of producing deep penetration welds, similar to those which are produced by electron beams. In the deep penetration mode, energy is delivered to the workpiece more rapidly than it can be removed by conduction and radiation. These conditions typically require power densities in the incident spot of the order of $10^6$ W/cm$^2$ ($\approx 10^7$ W/in$^2$). This results in the establishment of a deep penetration cavity, which is essentially a vapor column drilled through the thickness being penetrated such that energy is deposited through the material in depth. With relative motion between the workpiece and the focussed beam, the cavity, which is surrounded by a liquid pool in equilibrium (Fig. 1a), is translated through the work to form a deep penetration (high depth-to-width ratio) weld (Fig. 1b). The characteristics of this process have been modeled after a moving unsteady cylindrical heat source by Tong & Giedt(14).

With the development of higher power, industrially suited $CO_2$ laser systems (15), the maximum single-pass penetration in steel has been increased to approximately 20 mm (0.80 in) at a laser power of 20 kW (16). Five centimeter (2 inch) penetration has been demonstrated at 77 kW with a gas dynamic system. A 3.8 cm (1.5 inch) weld formed at 90 kW and a welding speed of 48 mm/sec (120 ipm) is shown in Fig. 2.

High power also facilitates welding of thin gage materials, at high speeds; .01 mm (.008 in.) material has been effectively welded at 1260 mm/sec (3000 ipm) at 6 kW. Typical welding performance at 5 kW in stainless steel involves welding speeds of 25 mm/sec (60 ipm) in 6.3 mm (1/4 in) thick material. Ten kilowatts is sufficient to weld 12.7 mm (1/2 in) thick material at the same speed (16, 18). The high welding speeds and the low specific energy inputs which result, as well as some unique characteristics of beam absorption have resulted in some unique weld properties (19-22) which will be described subsequently.

In the sections to follow, the capability and performance of laser welding are reviewed, along with the characteristics and properties of typical laser welds. A brief review of industrial applications is also presented.

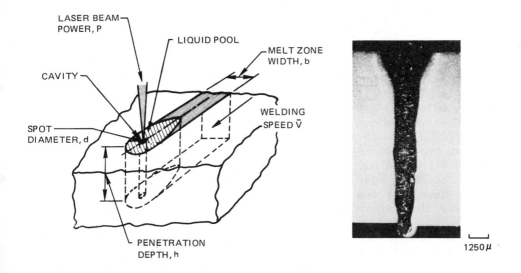

A. SCHEMATIC OF PENETRATION CAVITY          B. TYPICAL WELD

Fig. 1  Deep Penetration Welding Characteristics.

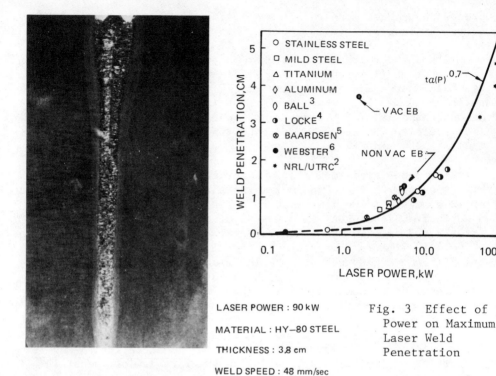

LASER POWER : 90 kW
MATERIAL : HY-80 STEEL
THICKNESS : 3.8 cm
WELD SPEED : 48 mm/sec

Fig. 2 Laser Weld Penetration in Stainless Steel.

Fig. 3 Effect of Power on Maximum Laser Weld Penetration

## LASER WELDING PERFORMANCE & CAPABILITY

It has been shown that the laser is capable of producing welds with high depth-to-width ratios, and that the specific energy per unit length of weld is substantially below that for conventional arc processes. Weld penetration as a function of power level in a variety of materials is shown in Fig. 3. Penetration increases approximately with the 0.7th power of the laser power level up to a maximum of 50 mm (2.0 in) at 77 kW. For comparison, Fig. 3 indicates that 20 mm penetration can be achieved with only 1.5 kW by an electron beam in vacuum. The maximum penetration per unit power level for the nonvacuum electron beam, however, appears to be much closer to that of the $CO_2$ laser.

The penetration attainable with the laser is related to total beam power and to the power density obtainable at the focal spot, which is related to the beam's specific focussing characteristics. At higher operating speeds, the laser compares favorably to the electron beam, however as speed is reduced in an attempt to increase penetration per unit power, the deep penetration cavity has a tendency to collapse due to the large mass of molten metal surrounding it, and the welds resemble fusion welds. The speed at which this breakdown of the deep penetration mode occurs is substantially higher than that for the electron beam because of the much shorter depth of field and the plasma attenuation problems encountered with the laser. Hence, a smaller maximum penetration is obtained.

A nondimensional correlation, initially developed by Hablanian (27) for electron beam welding has also been usefully applied to the laser as an assessment of its welding performance. Using dimensional analysis, the depth of penetration, h, may be expressed as a function of the laser beam power P and the diameter, d, of the focal spot, as well as the thermal properties of the material being welded: thermal conductivity, $\bar{K}$, thermal diffusivity $\alpha$, and an adjusted melting temperature $\bar{T}_m$ which incorporates the heat of fusion of the material. A process variable, the welding speed V, is also a part of the equation, which is:

$$\frac{h\bar{K}\bar{T}_m}{P} = f\,\frac{Vd}{\alpha}$$

This dimensionless correlation has been studied for laser welding (28) and is shown for a variety of pure metals and stainless steel in Fig. 4. It can be seen from Fig. 4 that laser welding performance is comparable to electron beam, provided that operation occurs well within the deep penetration mode.

Another means for evaluating laser welding performance is to examine the efficiency with which beam energy is utilized for fusion of material within the weld zone. Since the initial reflectivity of most metals for the 10.6 micron infrared wavelength of the $CO_2$ laser is generally quite high, it is necessary to determine what fraction of the beam energy is actually absorbed during the welding process, and how effectively this energy is then utilized to produce fusion. Results of calorimetric measurements of beam absorption and quantitative analysis of the efficiency of utilization of absorbed energy from (28) are summarized in Fig. 5. It can be seen that for operation within the deep penetration mode, energy absorption efficiency ranges from 55 to greater than 90 percent.

Melting efficiency is defined as the ratio of the total energy required to melt the material in the fusion zone to the actual energy absorbed. Welding penetrations on which these measurements were based were made at a laser power of 5kW and at a welding speed of 16.9 mm/sec (40 in/min). Melting efficiencies for elements studied ranged from 24 to greater than 70 percent, with the higher efficiencies being logically obtained in materials of low thermal diffusivity. Some of the measured values exceeded an analytically estimated maximum efficiency of 48.3%. If, however, the relationship in Ref. 29 is modified to include the heat of fusion, Hf, the expression

$$\eta_{melt} = \frac{0.483}{1-H_f/H}$$

is obtained. In light of the fact that the heat of fusion ($H_f$) of most metals is 20-30% of the total energy difference between the room temperature state and the liquid state at the melting temperature (H), the above relationship yields maximum values in the 70% range as observed experimentally.

It is also notable that welding at higher speeds, or more exactly, higher values of $V/\alpha$, should increase melting efficiencies for higher diffusivity materials as well. In the final analysis, the $CO_2$ laser is a highly efficient welding tool which when utilized within optimum operational ranges, provides the capability for forming welded joints with minimal energy input to the workpiece.

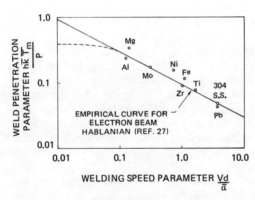

Fig. 4 Laser Welding Performance Correlation.

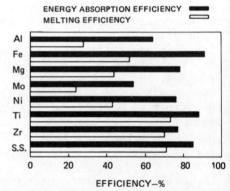

Fig. 5 Laser Welding Energy Absorption and Melting Efficiencies.

## LASER WELD CHARACTERISTICS AND PROPERTIES

### Ferrous Alloys

Typical deep penetration welds in ferrous alloys include welds in ship construction alloys (30) HY-130 alloy (20), and X-80 arctic pipeline steel (19). Grades A, B, and C ship steel have been successfully welded in thicknesses from 0.95 cm (3/8 in) (grade A) to 2.86 cm (1.125 in) (grade C). These steels contain 0.23% carbon (max.) and 0.60-1.03% Mn, with the amount of deoxidation increasing from grades A thru C. Welding tests in these alloys, were accomplished at power levels to 12.8 kW, making it necessary to weld the materials thicker than 1.90 cm in two passes, one from each side. Typical weld cross sections are shown in Figs. 6 and 7. Fig. 6 shows the cross section of two sided weld in 2.54 cm. thick material, along with a photo of the bead and a radiograph of a butt weld. Fig. 7 shows an autogenous welded "T" stiffener with the cross section of the weld in the inset. The web and flange of the "T" were 0.95 cm and 1.27 cm thick, respectively, and the weld was fabricated without filler metal by impinging the laser beam at a shallow (~ 7 1/2 deg) angle to the base plate.

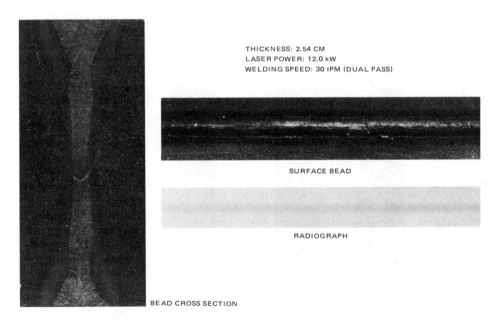

Fig. 6 Laser Weld Characteristics in Grade B. Ship Steel.

MATERIAL: GRADE B SHIP STEEL
THICKNESS: 1.05 & 1.27 CM
LASER POWER: 7.5 kW
WELDING SPEED: 2.54 CM/SEC

Fig. 7   Laser Tee Weld in Ship Steel.

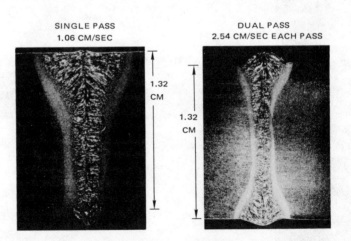

Fig. 8   Laser Welds in X-80 Arctic Pipeline Steel.

Material properties of the welds obtained in this study are summarized in Table I. All welds made within the thickness capability of the equipment exhibited good x-ray characteristics except for some root porosity in deep welds for which penetration capability was marginal. The generation of root porosity in dual pass welds was related to gas evolution without sufficient time for removal at the root of the second, or "blind" weld pass. Some reduction in porosity was attained by appropriate selection of welding parameters. Tensile properties of welds in all three grades were excellent, with failures occurring in the base plate rather than the welds. Adequate ductility was demonstrated by side bend tests.

Another example of laser welding of ferrous base alloys is the autogenous single and dual pass laser welds in 1.32 cm (0.520 in) thick X-80 arctic pipeline steel. Representative welds are shown in Fig. 8, and the production and evaluation of these welds is fully described in Ref. 19.

X-80 is an experimental pipeline alloy with 0.06 wt% C, 1.35% Mn, 0.6% Si, 0.65% Al, 0.1% Cb, 0.002% S, 0.011% N and 0.025% Ce. The laser welds in this alloy, made under the conditions specified in Fig. 8 exhibited characteristic columnar grains growing toward the cold base plate, in the direction of the maximum thermal gradient. Cross-weld tensile tests in this material, however, resulted in failures entirely in the base metal. The welds also exhibited excellent ductility in root, face and side-bend tests, with no cracks in these specimens observable by Magnaflux inspection.

Impact test results for base metal X-80, submerged arc welds, and both single and dual pass laser welds are summarized in Fig. 9. It was observed that, in contrast to the base metal which had 21°C (70°F) impact strength of 245J (181 ft-lb), both the single and dual-pass laser welds exhibited an upper Charpy shelf at this temperature which was in excess of the 358 J (264 ft-lbs) capacity of the impact tester. The base metal, however, showed a gradual linear decrease in Charpy energy down to a value of 122J (90 ft-lb) at -73°C (-100°F), while the single pass laser welds exhibited a sharp ductile-to-brittle transition to a lower Charpy shelf of approximately 13.6J (10 ft-lb) in the temperature range centering around -9°C (15°F). Metallographic examination of these specimens revealed a decrease in inclusion content which was felt to be responsible for the increase in upper Charpy shelf energy. (This effect, termed Fusion Zone Purification will be discussed under Metallurgical Characteristics below, using HY-130 as an example). In addition, a relatively coarse grain size (ASTM #3) was observed in the weld fusion zone, compared to ASTM #8 in the base plate. This

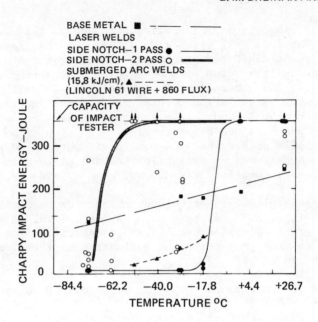

Fig. 9 Impact test Results for X-80 Laser and Submerged Arc Welds.

coarse grain size was considered to be responsible for the high ductile-to-brittle transition temperature.

Subsequently, dual-pass welds such as that shown in Fig. 8 were fabricated in order to reduce the energy input per unit length on each pass. This has the effect of increasing the cooling rate, which results in a finer grain size. A grain size reduction to ASTM #4 was achieved, and the ductile-to-brittle transition temperature of these welds was reduced to below -51°C (-60°F). Comparison data for submerged arc welds in the same alloy also appear in Fig. 9. The best of a large series of submerged arc welds were formed with Lincoln 61 wire and No. 860 flux, but did not approach base metal impact properties.

Regarding other ferrous materials, satisfactory laser welds have been made in a variety of low carbon alloys. Typically, "hardenability-related" problems as a result of cooling rate do not occur in alloys of 0.15% C or lower. Yessick and Schmatz (31) have indicated the ability to obtain satisfactory welds in HSLA grade steels as well, citing properties equivalent to base metal. They note that the low specific energy required for laser welding of HSLA grades minimizes strength decreases stemming from the high energy inputs of arc processes.

## Titanium Alloys

A study comparing electron beam, laser, and plasma/gas tungsten-arc welds in Ti-6Al-4V up to 0.64 cm (0.25 in) thick (32) has been conducted in which butt weld specimens in a variety of thicknesses were prepared and subjected to metallographic analysis and destructive & nondestructive tests. Laser welds were found to be radiographically sound & exhibited tensile strengths in excess of the base material. Due to the low specific energy input (high cooling rate) the laser welds were fine-grained and somewhat harder than the base metal. As a result of the fine acircular martensitic structure in the fusion zone, of these rapidly-chilled welds, it was evident that post weld processing would be necessary for applications requiring high fracture toughness.

The fatigue properties of autogenous laser butt welds were investigated in another study (22) for thicknesses of 0.36 and 0.58 cm (0.14 and 0.23 in). Typical welds for these materials are shown for a variety of speeds in Fig. 10. The mechanical properties and tendencies for porosity were investigated, for a variety of welding conditions. All tensile specimens fialed in the base metal, and the welds did not exhibit excessive oxygen contamination when properly shielded. A high resolution radiography technique, capable of

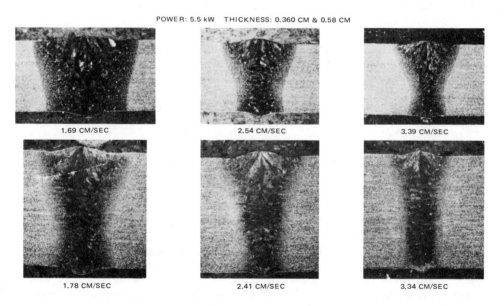

Fig. 10 Macrographs of Laser Welded Titanium - 6 Aluminum - 4 Vanadium Alloy.

resolving porosity as small as 0.015 cm (0.006 in) was used to inspect the welds, and with the exception of one specimen, no detectable porosity was found.

The fatigue test results on all laser welds compared favorably with the best welds made previously by a special plasma-arc technique. The results indicated that Ti-6Al-4V laser welds can be made with fatigue properties comparable to the base metal. The best laser weld fatigue specimens exhibited failure initiation sites in the base metal rather than in the welds. Other weld failures initiated at very small pores, below the limit of resolution of the radiographic inspection. Though these failures represented decreases in fatigue endurance below that of the base metal, it was concluded that laser welding exhibited a minimal tendency for formation of small pores or voids thus the laser is a potentially useful means for welding thin-to-medium sections of Ti-6Al-4V.

## Aluminum Alloys

Aluminum alloys, as a group, have been the most difficult to laser weld of the commonly used structural alloys. This difficulty stems from the high initial surface reflectivity of aluminum for $CO_2$ laser radiation, combined with the high thermal diffusivity of the alloy. The most successful welding of aluminum alloys has been fillet welding. Although considerable porosity often results, weld mechanical properties are normally acceptable.

Some success has been achieved in the butt welding of 2219 & 5456 alloys. Both of these materials have been laser welded up to 0.95 cm (0.375 in) thickness of acceptable bead profiles and reasonably low levels of porosity. Tensile tests of welds in these alloys resulted in failure by diagonal shear through the fusion zone at >99% of parent metal strength. The fractures did not appear to be directly related to the porosity present in the welds.

A major problem still remaining to be solved in the butt welding of aluminum alloys is the difficulty of fabricating acceptable butt welds reproducibly. Very often, once full penetration is established, excessive bead drop-through occurs. Anything short of full penetration (blind welds) results in excessive and unacceptable levels of root porosity. The reproducibility problems appear to be related to the delicate balance between initiation of efficient coupling, and the maintenance of adequate penetration, without overpenetration. These problems have not yet been solved, and it is clear that it will be necessary to control process parameters more closely than has been found to be necessary with ferrous, titanium and nickel base alloys. With improved process development, the capability for reproducible butt welding of aluminum alloys is expected to improve.

## High Temperature Alloys

$CO_2$ lasers have been used successfully to weld a variety of superalloys, including PK 33, N 75 and C 263 in thicknesses of 1.0 and 2.0 mm. (0.039 and 0.079 in) (33). A high temperature ferrous base alloy, Jethete M152 was also successfully welded in this study. It was reported that 14 of 18 tensile specimens failed in the base material, away from the weld zone. The Nimonic 75 alloy, however, exhibited tensile fractures predominantly located at the edge of the fusion zone. Fracture occurred at 90% of parent material strength, with only 12-13% elongation as compared to 33-35% in the parent material. As expected, welds in the Jethete M-152 material exhibited higher hardness, than the base metal (480 DPN vs 300). The nickel base alloy tensile specimens failed with ductile cup-cone fractures.

It was found, for welding of Ni base superalloys, that an energy density threshold existed below which the workpiece reflectivity was high and prevented useful welding. This has also been found for other metals & alloys. For superalloys, it was found that a low order mode laser was capable of exceeding the threshold, entering a region of low reflectivity in which the laser energy was efficiently absorbed producing welds. It was concluded that a low-order-mode $CO_2$ laser could weld a significant fraction of the materials used in aircraft engine fabrication. The temperature resistant nickel and iron base alloys appeared to demonstrate good laser weldability, and exhibited useful weld mechanical properties within the thickness range investigated.

## Welds Utilizing Filler Metal Addition

A practical consideration in the application of laser welding is the ability to add filler material to the fusion zone. This requirement stems both from nonperfect joint fitup and from the need for modification of fusion zone metallurgical characteristics to attain desired weld properties. In spite of the relatively narrow molten zone associated with laser welding, continuous addition of filler material in wire form (approximately 1 mm diameter) has been demonstrated. Using filler material, sound weldments have been formed between 1 cm thick alloy steel plates separated by a 1.5 mm gap. The general nature of the filler addition process is shown in the comparison in Fig. 11. At the left is shown a butt weld formed without filler which exhibits a substantial underfill due to the vee groove in the weld joint preparation. With filler addition a sound weldment with good top and bottom bead reinforcement is attained. Somewhat surprising, perhaps is the fact that the weld with filler addition is actually narrower than the one without filler. This is due to the fact that the weld energy input per unit length was the same in both instances illustrated in Fig. 11. With filler addition, some of the beam energy was required to melt the filler so that less energy was available to melt parent material; the total volume of fused material is essentially the same in both cases.

It is noted that the general characteristics of the deep-penetration weld are retained with filler addition provided that the filler metal is not added so rapidly that it results in collapse of the deep-penetration cavity. This behavior is quite significant in that it indicates that effective control of weld metallurgy and limited compensation for imperfect fitup may be attained without significant loss of weld performance. That is, the laser weld profile is essentially the same at constant specific energy input regardless of whether a bead-on-plate penetration, a tight butt weld or a gapped butt weld with filler addition is formed.

## UNIQUE METALLURGICAL CHARACTERISTIC OF LASER WELDS

### Fusion Zone Purification

One of the most interesting developments in the field of laser welding to date was the discovery that the $CO_2$ laser is capable of making autogenous welds in a variety of ferrous base materials while simultaneously improving the microstructure of the fusion zone. Welds have been produced with mechanical properties which are at least equivalent to, and often substantially better than those of the original alloy. This behavior has been described in Refs. 19-21.

As an example of this phenomenon, welds in three thicknesses of HY-130 alloy steel, representative of the tests made in one study, are shown in Fig. 12. These welds were subjected to a variety of mechanical tests, as well as to metallographic examination. All specimens easily withstood bends to 3t radius with no cracking as determined by Magnaflux. Hardness in the fusion zone was $R_c$ 39.8, somewhat higher than the base material (34.4 $R_c$). The highest hardness up to $R_c$ 43.2, were found in the HAZ. As expected considering the correspondence between strength and hardness in ferrous alloys no tensile specimens failed in the fusion zone. Dynamic tear tests of the weld metal revealed DT energies which ranged, depending on specimen thickness, from the higher end of the normal base metal range for HY-130 steel to values considerably in excess of the highest base metal values as shown on the RAD diagram. Figs. 13 and 14 are indicative of the "best case" and "worst case" welds for the 1.27 cm (0.50 in) welds, with the best case being the weld with the fewest inclusions and the worst case being the weld with the largest number of inclusions. Welds in all thicknesses clearly demonstrated

# WELDING WITH HIGH POWER LASERS

marked reductions in the resolvable weld inclusion contents (10 μ resolution limit) as compared with an immediately adjacent equal area of base metal. In specimen C-1 only 24.5 percent of base metal inclusions remained in the weld metal while in specimen C-7, only 18 percent were retained. Inclusions were shown to be aluminum oxides with small amounts of calcium, except for rare-earth treated heats which showed predominantly lanthanum/cerium oxide inclusions. Reduction of inclusion content in alloy steels has been shown (34) to be a dominant factor in improving the resistance of these alloys to impact-type fractures.

The ability of the $CO_2$ laser to purify a fusion zone stems from the fact that, in the deep penetration mode the metal is exposed to laser radiation throughout its thickness. This coupled with the greater absorptivity of nonmetallic inclusions for 10.6μ radiation promotes preferential fusion and vaporization of the inclusions which in vapor form can then be aspirated from the melt pool through the deep penetration cavity. It has been shown by chemical and fracture surface analysis that inclusions are actually removed, not just dissolved or refined, during this fusion zone purification process.

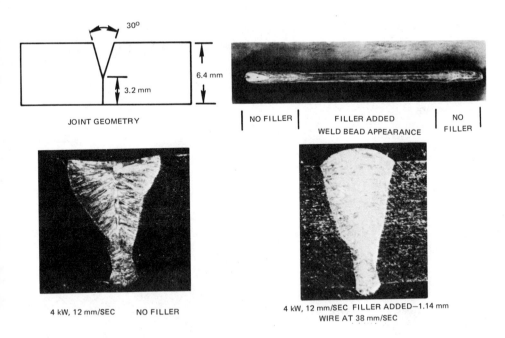

Fig. 11 Laser Weld Filler Metal Characteristics.

0.64 cm THICK

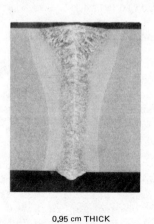
0.95 cm THICK

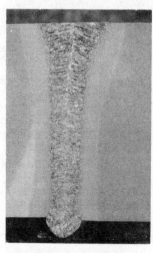
1.27 cm THICK

Fig. 12  Laser Welds in Three Section Thicknesses of HY-130 Alloy Steel

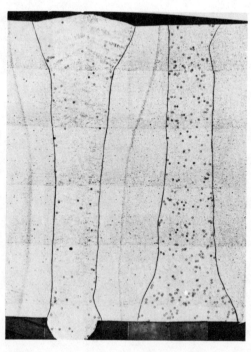

WELD FUSION ZONE—74 INCLUSIONS
BASE METAL—302 INCLUSIONS

(NOTE: REDUCTION IN MAGNIFICATION FROM ORIGINAL PHOTOGRAPHIC MAP REQUIRED ENCIRCLING OF INCLUSIONS FOR CLARITY)

100μ

Fig. 13  Distribution of Inclusions in the Weld Fusion Zone and Equivalent Area of Base Metal in HY-130 Laser Weld.

# WELDING WITH HIGH POWER LASERS

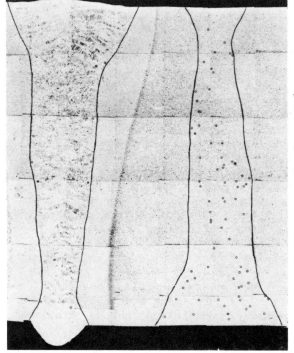

WELD FUSION ZONE—12 INCLUSIONS
BASE METAL—67 INCLUSIONS

(NOTE: REDUCTION IN MAGNIFICATION FROM ORIGINAL PHOTOGRAPHIC MAP REQUIRED ENCIRCLING OF INCLUSIONS FOR CLARITY.)

Fig. 14 Distribution of Inclusions in the Weld Fusion Zone and Equivalent Area of Base Metal in HY-130 Laser Weld.

$10\mu$

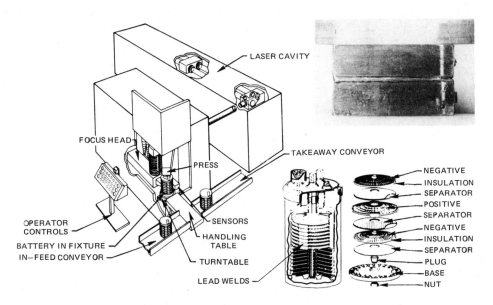

Fig. 15 Laser System for Welding of Storage Batteries (insert shows weld-lead battery lug).

## Industrial Welding Applications

Although substantial advances have been made in laser welding technology within the past several years, only two multikilowatt systems are currently employed in production welding. These units are being used from butt welds between positive lugs of a cyclindrically-shaped, lead-acid storage battery Fig. 15. In production, the battery stack is assembled from individual layers and fixtured on a rotating table within a safety enclosure which incorporates an exhaust for the lead fumes. The system locates the seam between adjacent layers and forms a partial penetration butt weld at a welding speed of approximately 850 mm/sec (200 ipm.). Because of the high welding speed, a relatively narrow fusion zone is generated which facilitates welding of the units in the horizontal position. These two systems are now in their third year of production operation and are demonstrating the reliability required for high volume production applications.

## Concluding Remarks

In view of the excellent weld characteristics obtained with lasers in many materials it may be surprising that only two production machines are in use today. One reason for the relatively slow adaptation rate is, of course, cost; laser systems are substantially more expensive than more conventional processing systems In spite of this high initial cost, however, lasers can be quite cost effective in many projected applications. Absence of reliability and actual production performance information has therefore been a principal factor in delaying industry acceptance. As high power industrial laser welding reliability information is generated by the initial applications, it is anticipated that the laser will find ever widening acceptance as a versatile, cost-effective tool for industrial welding.

## References

1. Maiman, T. H.: Stimulated Optical Radiation in Ruby Masers. Nature, Vol. 187, 1960, p. 493.

2. Charschan, S. S. Editor: Lasers in Industry. Van Nostrand Reinhold Co., New York, 1972.

3. Ready, J. F.: Effects of High Power Laser Radiation. Academic Press, New York, 1971.

4. Geusic, J. E., et al.: Laser Oscillations in Nd-Doped Yttrium Aluminum, Yttrium Gallium, and Gadolinium Garnets, Applied Physics Letters, Vol. 4, 1964, p. 182.

5. Patel, C. K. N.: Continuous-Wave Laser Action on Vibrational-Rotational Transitions of $CO_2$. Physical Review, Vol. 136, 1964, p. A1187.

6. Patel, C. K. N., P. K. Tien and J. H. McFee: CW, High-Power $CO_2$-$N_2$He Laser. Applied Physics Letters, Vol. 7, No. 11, Dec. 1965, pp. 290-292.

7. Harrigan, F. A., R. I. Rudko and D. T. Wilson: High-Power Gas Laser Research. Raytheon Report S-1037 on Contract DAAH01-67-C1589, Jan. 10, 1968.

8. Brown, C. O.: High Power, $CO_2$, Electric Discharge Mixing Laser. Applied Physics Letters, Vol. 17, No. 9, Nov. 1970, pp. 388-391.

9. Davis, J. W. and C. O. Brown: Electric Discharge Convection Lasers. Paper 72-722 presented at the AIAA 5th Flux and Plasma Dynamics Conference, Boston, MA, June 1972.

10. Alwang, W. G., L. A. Cavanaugh, and E. Sammartino: Continuous Butt Welding Using a Carbon-Dioxide Laser. Welding Research Supplement, Mar. 1969, pp. 1105-115S.

11. Banas, C. M.: The Role of the Laser in Materials Processing. Presented at the Canadian Welding and Materials Technology Conference, Toronto, Canada, Sept. 29 - Oct. 1, 1969.

12. $CO_2$ Laser Welding Joins the Parade: Welding Engineer, Aug. 1970, pp. 42-44.

13. Brown, C. O. and C. M. Banas: Deep-Penetration Laser Welding. Paper presented at the AWS 52nd Annual Meeting, San Francisco, CA, April 26-29, 1971.

14. Tong, H. and W. H. Giedt: Depth of Penetration During Electron Beam Welding. ASME Paper No. 70-WA/HT-2.

15. Burwell, W. G.: Review of High Power Laser Technology. Presented at the 3rd R.P.I. Workshop on Laser Interactions and Related Plasma Phenomena, Troy, NY, Aug. 13-17, 1973.

16. Locke, E. V., E. D. Hoag and R. A. Hella: Deep-Penetration Welding with High Power, $CO_2$ Lasers. IEEE Journal of Quantum Electronics, Vol. QE-8, No. 2, Feb. 1972.

17. Banas, C. M.: Laser Welding Developments. Proceedings of the CEGB International Conference on Welding Research Related to Power Plant, Southampton, England, Sept. 17-21, 1972.

18. Ball, W. C., and C. M. Banas: Welding with a High-Power, $CO_2$ Laser, SAE Paper No. 740863, Oct. 1974.

19. Breinan, E. M. and C. M. Banas: Preliminary Evaluation of Laser Welding of X-80 Arctic Pipeline Steel. WRC Bulletin No. 201, Dec. 1974.

20. Breinan, E. M. and C. M. Banas: Fusion Zone Purification During Welding with High Power $CO_2$ Lasers, Proceedings of the Second International Symposium of the Japan Welding Society, Osaka, Japan, Aug. 25-29, 1975.

21. Breinan, E. M., C. M. Banas, G. P. McCarthy and B. A. Jacob: Evaluation of Basic Laser Welding Capabilities, Technical Report R77-911989-10, Contract N00014-74-C-0423, United Technologies Research Center, E. Hartford, Conn., March 1977.

22. Breinan, E. M. and C. M. Banas: Fatigue of Laser-Welded Titanium 6Al-4V Alloy, Report No. R75-412260-1, United Technologies Research Center, East Hartford, CT, July 1975.

23. Baardsen, E. L., D. J. Schmatz and R. E. Bisaro: High Speed Welding of Sheet Steel with a $CO_2$ Laser, Welding Journal, April 1973.

24. Webster, J. W.: Welding at High Speed with the $CO_2$ Laser, Metal Process, Nov. 1970.

25. Meier, J. W.: High Power Density Electron Beam Welding of Several Materials. Paper presented at the Second International Vacuum Congress, Washington, D.C., Oct. 10-16, 1961.

26. Meier, J. W.: Recent Developments in Non-vacuum Electron Beam Welding. Paper presented at the International Conference on Electron and Ion Beam Science and Technology, Toronto, Canada, May 6, 1974.

27. Hablanian, M. H.: A Correlation of Welding Variables. Proceedings of the 5th Symposium on Electron Beam Technology, 1963.

28. Breinan, E. M. and C. M. Banas: Evaluation of Basic Laser Welding Capabilities, Technical Report R76-911989-4, Contract N00014-74-C-0423, United Technologies Research Center, East Hartford, Conn., Nov. 1975.

29. Swifthook, D. and A. E. F. Gick: Penetration Welding with the Laser, AWS Journal, Nov. 1973.

30. Banas, C. M. and G. T. Peters: Study of the Feasibility of Laser Welding in Merchant Ship Construction. Final Report to Bethlehem Steel Corp. in support of Contract 2-36214-U.S. Department of Commerce, Aug. 1974.

31. Yessick, M. and D. J. Schmatz: Laser Processing in the Automotive Industry. SME Paper MR74-962, 1974.

32. Banas, C. M., D. M. Royster, D. Rutz, and D. Anderson: Comparison of Electron Beam, Laser Beam, and Plasma-Arc Welding of Ti-6Al-4V. Presented at the 56th Annual Meeting of the American Welding Society, Cleveland, Ohio, Apr. 21-25, 1975.

33. Adams, M. J.: $CO_2$ Laser Welding of Aero Engine Materials. Report No. 3335/3773, British Welding Institute, Cambridge, England, 1973.

34. Hill, D. C., and D. E. Passoja, "Understanding the Role of Inclusions and Microstructure in Ductile Fracture," Welding Research Supplement to the Welding Journal, Nov. 1974, pp. 481S-485S.

35. Seaman, F. D., "Welding with a Multikilowatt $CO_2$ Laser," SME Paper No. MR74-957.

# FUNDAMENTALS OF SUPERPLASTICITY AND ITS APPLICATION

O. D. Sherby[1], R. D. Caligiuri[2], E. S. Kayali[3], and R. A. White[4]

(1) Professor, Stanford University, Stanford, Ca. 94305
(2) Materials Scientist, SRI International
    Menlo Park, Ca. 94025
(3) Asst. Prof., Istanbul Tech. Univ., Istanbul, Turkey
(4) Manager, Matls. and Corr., Bechtel National, Inc.
    San Francisco, Ca.

## Abstract

The term superplasticity has been used to describe extraordinary elongations (several hundred percent) obtained during tensile deformation of polycrystalline materials. Mostly a scientific curiosity fifteen years ago, superplastic materials are now being used in a number of industrial applications, including near-net-shape forming and solid state welding. Two types of superplastic flow have been observed: internal stress superplasticity and fine structure superplasticity. Both rely on one common characteristic: a high sensitivity of the flow stress to the strain rate. This is a necessary but not sufficient condition for superplasticity. The various structural prerequisites for fine structure superplasticity are evolving rapidly and the phenomenology of superplastic flow is well documented. This knowledge, coupled with advances made in understanding normal plastic flow in crystalline solids (diffusion controlled dislocation creep), permits predicting methods for enhancing and optimizing superplasticity in materials.

## Introduction

Superplasticity refers to the ability of certain polycrystalline metallic materials to extend plastically to large strains (e > 400%) when deformed in tension. In general superplastic materials exhibit low resistance to plastic flow in specific temperature and strain rate regions. These characteristics of high plasticity and low strength are ideal for the manufacturer who oftentimes needs to fabricate a material into a complex but sound body with a minimum expenditure of energy.

## Application of Superplastic Properties

An example of the superplastic forming of a nickel base alloy is shown in Figure 1. This work was performed at Pratt and Whitney in Florida[1]. The sample on the left is a billet of IN-100 (60Ni, 10Cr, 15Co, 3Mo, 5.5Al, 4.5Ti, .18C and 1V) prepared by powder metallurgy processing. This billet was inserted into a die and warm pressed to the configuration shown in the center of Figure 1. This disk shaped object was then placed into another die and superplastically pressed into the complex shaped component shown in Figure 1. This final product is a single integrated component consisting of a disk with turbine blades. This product is virtually ready for use in a jet engine for elevated temperature service. Only some minor machining is performed on the disk, followed by selective heat treatment of the turbine blades. The latter treatment leads to grain coarsening, thus enhancing the creep resistance of the IN-100 turbine blades. Another example of superplastic shaping, performed at Rockwell International[2], is shown in Figure 2. The particular component is a nacelle beam (part of an aircraft frame) which previously was made by fastening and joining 104 separate components. This rather complex structure can be made by a single forming operation using one monolithic piece of a superplastic titanium alloy. Superplastic forming, in this case, led to a weight saving of 33% and cost saving of 55% compared to the prior art of joining and fastening the many separate components. The examples given in Figures 1 and 2 represent the primary unique characteristic of superplastic materials. That is, such materials can be superplastically formed and expensive machining, joining or fastening operations can be avoided. For components that are difficult to machine, this novel processing method can lead to significant cost savings.

A second important advantage of superplastic materials is that whereas they are weak at forming temperatures (0.4 - 0.7$T_m$, where $T_m$ is the melting temperature in degrees absolute), they are generally very strong and ductile at low temperatures. This characteristic is related to the fine grained structure associated with most superplastic materials. Figure 3 gives an example[3] showing the low strength exhibited by a fine grained Al-Zn alloy at warm temperature (0.7$T_m$) where it is superplastic, but, at low homologous temperatures, this fine grained alloy exhibits high strength. Oftentimes, superplastic materials can be heat treated after superplastic forming to yield unusual properties. Mention was already made regarding grain coarsening for enhancing high temperature properties in nickel-base superalloys. Heat treatment for precipitation and/or transformation hardening is possible with superplastic titanium alloys, superplastic aluminum alloys and superplastic ultrahigh carbon steels. Such treatments can also lead to unique mechanical properties such as high strength and ductility and good wear resistance.

# FUNDAMENTALS OF SUPERPLASTICITY

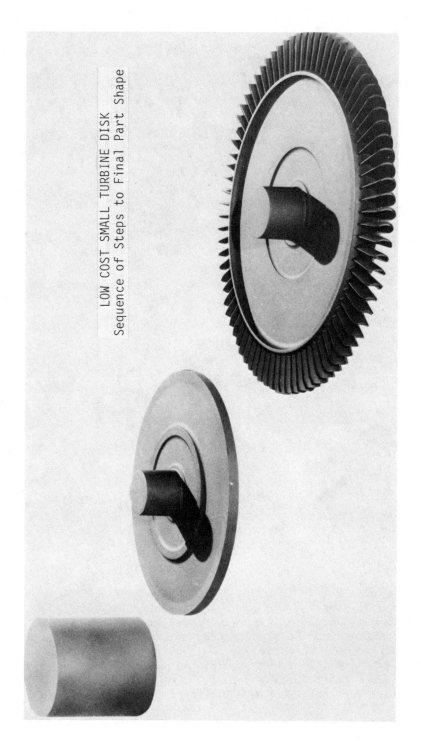

LOW COST SMALL TURBINE DISK
Sequence of Steps to Final Part Shape

Figure 1. Net shape forming of an ultrafine grain size nickel base alloy by superplastic forming in two stages. A. Original powder metallurgy IN 100 billet. B. Powder metallurgy billet pressed into disk shape. C. Disk shape billet superplastically pressed into disk and turbine blades. Courtesy of J. Moore and R. Athey, Pratt and Whitney, Florida. (AF Contract F33615-72-C-2177)

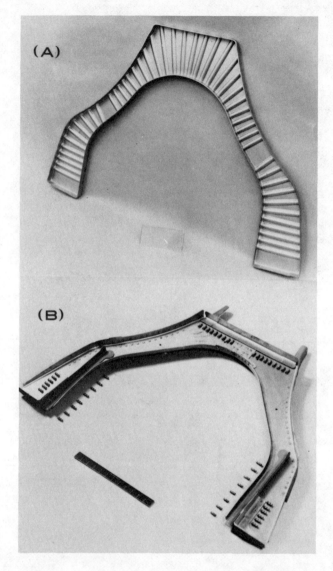

Figure 2. The nacelle beam frame shown in (A) above was made in one superplastic forming operation. It was formed from a Ti-6 Al-4V alloy at a temperature around 920°C. Conventional methods to make this part required the preparation of 104 separate components which were then fastened and joined by conventional techniques; the separate components are illustrated in (B) above. Courtesy of H. Hamilton and E. Weisert, Rockwell International, Thousand Oaks, California.

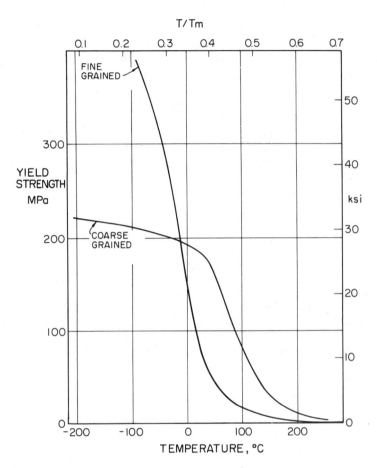

Figure 3. The influence of grain size on the strength of a monotectoid composition Al-Zn alloy (78Zn - 22Al) is shown as a function of temperature. At high temperatures ($\sim 0.6 T_m$) the fine grain Al-Zn alloy is weak and superplastic whereas at low temperatures ($\sim 0.25 T_m$) it is strong.

A third unique characteristic that can be associated with fine grained superplastic materials is their ability to bond readily in the solid state. These materials can be solid state pressure bonded to each other, or to other similar materials, at moderately low temperatures where they are superplastic (i.e., superplastic "glue"). Diffusion bonding occurs readily because of the high volume fraction of grain boundaries which act as short circuit paths and accelerate diffusion. The relative ease of plastic flow at the interface also contributes to the bonding process. Pressure bonding of superplastic metallic powders is another area of technology where novel structures and properties can be developed.

## Applied mechanics aspects of superplasticity

Superplastic metallic alloys have one primary property in common. The strength (resistance to plastic flow) of these materials is highly strain rate sensitive. That is, the flow stress, $\sigma$, increases rapidly with strain rate, $\dot{\varepsilon}$. A measure of this rate sensitivity of strength is given by the relation

$$\sigma = K\dot{\varepsilon}^m \qquad (1)$$

where K is a material constant and m is the strain rate sensitivity exponent.* The value of m determines the rate at which the neck progresses after localized plastic flow starts. As m increases, the elongation to fracture increases[4]. This experimental fact is shown in Figure 4 which illustrates the relationship between typical elongations to fracture and the corresponding m value for several materials[5-9] tested at $0.7T_m$. The elongation to fracture is seen to increase to large values as m approaches about 0.5. When the relation between $\sigma$ and $\dot{\varepsilon}$ is linear (m = 1), the material is said to behave in a Newtonian-viscous manner; hot glass, tar, well-masticated chewing gum all obey the Newtonian-viscous relationship, and these materials can be classified as ideally superplastic.

Applied mechanics can describe the profile change of a tensile sample with strain once a neck forms, provided the strain rate sensitivity exponent is known. When m is low, the increase in stress at the neck will lead to a large increase in strain rate in that region. Thus, the neck will grow sharply, leading to premature

---

* Oftentimes the stress exponent, n, is used to describe the relation between the strain rate (creep rate) and the flow stress. In this case, the expression, $\dot{\varepsilon} = K'\sigma^n$, is used, where K' is a material constant. The stress exponent, n, is equal to $\frac{1}{m}$, where m is the strain rate sensitivity exponent m [equation (1)]. When data are plotted as logarithm $\sigma$ versus logarithm $\dot{\varepsilon}$, the resulting slope is equal to m. When data are plotted as logarithm $\dot{\varepsilon}$ versus logarithm $\sigma$, the result slope is equal to n. In this paper, we will use m and n interchangeably.

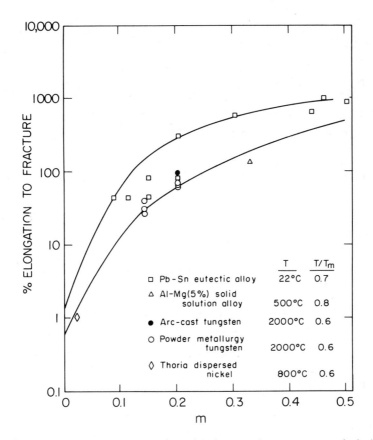

Figure 4. Metals and alloys show high strain rate sensitivity at high temperatures. This rate sensitivity is usually measured by the exponent m in the phenomenological equation $\sigma = K\dot{\varepsilon}^m$. The above figure shows that the ductility (measured by elongation to fracture) increases as the strain rate sensitivity exponent increases. Superplastic metals and alloys generally exhibit strain rate sensitivity exponent (m) values in the range m ≃ 0.5 (0.4 to 0.6). Data from references 5 to 9.

failure and a low elongation to fracture. Conversely, when m is large, the strain rate increases slowly from the increased stress in the neck region. As a result, the neck forms gradually. Various methods of applied mechanics ( a macroscopic approach) have been applied to predict the elongation to fracture of materials that fail by necking. Four of these are outlined below:

(a) Rossard (10) has shown that the strain at the start of necking is given by $\varepsilon_{neck} = \frac{N}{1-2m}$ where N is the strain hardening exponent in $\sigma = K'\varepsilon^N$. His theory would predict infinite plasticity at m = 0.5 provided N has a finite positive value.

(b) Morrison (11) indicated that the dimensions of the sample will dictate the total elongation observed and showed that

$$\% \text{ elongation to fracture} = bm^2 \left(\frac{d_o}{L_o}\right) \times 100$$

where b is a material constant, and $d_o$ and $L_o$ are the initial diameter and length of sample, respectively.

(c) Avery and Stuart (12) took into account the possible tapered shape of the sample. Their equation is given as

$$\% \text{ elongation} = \left\{\left[\frac{1-\beta^{1/m}}{1-\alpha^{1/m}}\right]\right\} \times 100$$

where $\alpha$ is the ratio of the minimum to maximum area at the start of the test and $\beta$ is the same ratio at some arbitrary stage of the test at which the elongation is measured.

(d) Burke and Nix (13), using a finite element approach, showed tha

$$\% \text{ elongation} = \left[\exp\left(\frac{2m}{1-m}\right)-1\right] \times 100.$$

Figure 5 illustrates the predictive ability of three of the four relations described above. As can be seen, the elongation to fracture for a number of high carbon and ultrahigh carbon steels[14, 15] agree well with the predictive relations based on a knowledge of the strain rate sensitivity exponent, m.

We thus see, from applied mechanics, that a high strain rate sensitivity exponent, m, leads to high ductility. When m ≅ 0.5, the material is generally superplastic (>400% elongation). A high value of m is a necessary condition for superplasticity. It is, however, not a sufficient condition. Material embrittling characteristics such as grain boundary separation, cavitation at interphase boundaries, and other premature failure modes can lead to early failure even when the material is highly strain rate sensitive. It is the knowledge embodied in the field of materials science that permits us to prepare the appropriate microstructure which incorporates all the major criteria needed for developing truly superplastic metallic alloys. Therefore, the remainder of our paper will emphasize the materials science aspects of superplasticity.

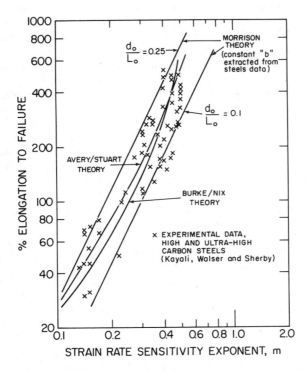

Figure 5. The predicted elongation to fracture is shown as a function of the strain rate sensitivity exponent, m, based on various analytical expressions (using an applied mechanics approach). The curves predicted by the Morrison relation were calculated using b = 100. The elongation and m value data are for high and ultrahigh carbon steels[14,15].

Materials Science Aspects of Superplasticity

There are two major types of superplasticity that have been considered in the past. One can be termed as "internal stress" superplasticity and the other as "fine structure" superplasticity. Evidence will be presented for each type of superplastic behavior and each will be discussed in terms of the phenomenology that leads to predictive relations for such materials.

(A) Internal Stress Superplasticity

The time dependent plastic flow behavior (creep) of crystalline metallic solids is well established at warm and high temperatures ($>0.4T_m$)[16-18]. If such materials are exposed to special conditions that create internal stress, the creep rate may be enhanced due to the excess plastic flow associated with the additive effect of the internal and external stresses. We cite two examples which illustrate internal stress superplasticity. One is the case of materials that exhibit large anisotropy in their thermal expansion coefficients ($\alpha$). Alpha uranium is a case in point; the anisotropy in expansion coefficient for uranium[19] is such that $\alpha$ in the [100] direction is positive and equal to $+ 45 \times 10^{-6}$ per °C, whereas $\alpha$ in the [010] direction is negative and equal to $-15 \times 10^{-6}$ per °C (measured at 500°C). Thermally cycling a non-textured sample of polycrystalline uranium in the alpha region will lead to high stresses at the grain boundaries of adjoining grains since each grain will expand or contract in a different direction. These internal stresses will assist the external stress in generating additional plastic flow. An example showing the creep behavior of thermally cycled alpha uranium compared to the creep behavior of the same material under isothermal conditions[20-24] is shown in Figure 6. The graph shows a plot of the diffusion compensated creep rate versus the modulus compensated stress. Three important observations can be noted: (1) thermally cycled uranium is much weaker than isothermally deformed uranium ($10^6$ difference in creep rate at $\sigma/E = 10^{-5}$), (2) thermally cycled uranium exhibits a strain rate sensitivity exponent at low stresses equal to unity, and (3) the thermally cycled samples and the isothermally tested samples yield data that converge at high stress; this is expected since the internal stress generated by thermal cycling (a constant) will have a diminishing contribution to creep as the applied stress is increased.

The high strain rate sensitivity of the thermally cycled sample leads to superplastic behavior at low stresses. Figure 7, taken from Lobb, Sykes and Johnson[20] illustrates this behavior for polycrystalline alpha uranium. Whereas the sample tested isothermally at 600°C exhibits about 50% elongation at fracture, the thermally cycled sample exhibits no sign of failure at the time testing was stopped (elongation of 500%). The high strain rate sensitivity of the thermally cycled sample shown in Figure 6 can be explained through the concept of an internal stress. In the

# FUNDAMENTALS OF SUPERPLASTICITY

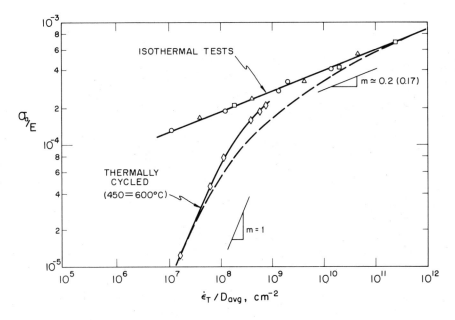

Figure 6. Comparison of the creep behavior of alpha uranium under isothermal conditions and under thermal cycling conditions. The strain rate sensitivity exponent is high under thermal cycling, and the material behaves superplastically. The predicted behavior of the thermally cycled sample, by introducing the concept of an internal stress [equation (4b) in text], is given by the dashed line. Note: In order to determine $\dot{\varepsilon}/D$, it is necessary to use a diffusion coefficient which represents the average diffusivity during the variable temperature-time cycle; this was done by graphical integration using the relation $D_{avg} = [D_o \int_0^{t_1} \exp(-Q_L/RT)dt]/t_1$, where $t_1$ is the time for a complete thermal cycle.

α - URANIUM

ISOTHERMAL
TENSILE TEST, 600°C (50% elong.)

UNDEFORMED SAMPLE

THERMALLY CYCLED
(450 ⇄ 600°C)   (>500% elong.)

Figure 7. Thermal cycling of alpha uranium leads to a highly strain rate sensitive material. This high strain rate sensitivity is manifested in high ductility as can be seen by the specimen at the bottom of the adjoining figure. Whereas creep deformation occurring under isothermal conditions leads to 50% elongation at 600°C, thermal cycling between 450°C - 600°C leads to creep elongations exceeding 500%. (From Reference 20)

# FUNDAMENTALS OF SUPERPLASTICITY

following we develop a creep model to predict the high strain rate sensitivity of materials where an internal stress exists. Certainly, a thermally cycled sample of polycrystalline, non-textured alpha uranium will have areas of internal stress which add to the applied stress and areas of internal stress which subtract from the applied stress, depending upon their orientation in relation to the direction of applied stress. Statistically, in a non-textured, polycrystalline sample, the magnitude of the average positive internal stress will be equal to the magnitude of the average negative internal stress: $|(\sigma_i)_{pos}| = |(\sigma_i)_{neg}| \equiv \sigma_i$. We assume that every area in the sample is subjected to either a positive or negative internal stress. We further assume that the deformation mechanisms active in one area of internal stress operate independent of the deformation mechanisms active in a neighboring area. Given the above conditions and assumptions, we can write;

$$\dot{\varepsilon}_T = C_1 (\dot{\varepsilon})_{\sigma_i = pos} + C_2 (\dot{\varepsilon})_{\sigma_i = neg} \tag{2}$$

where $\dot{\varepsilon}_T$ = the total creep rate, $(\dot{\varepsilon})_{\sigma_i = pos}$ = the contribution to the creep rate from areas of positive internal stress, $(\dot{\varepsilon})_{\sigma_i = neg}$ = the contribution to the creep rate from areas of negative internal stress, and $C_1$ and $C_2$ are constants describing the relative amounts of positive and negative internal stress. The condition of zero deformation at zero applied stress in a polycrystalline, non-textured sample demands that the total amount of positive internal stress equal the total amount of negative internal stress. Hence, $C_1 = C_2 = 1/2$ and we can write:

$$\dot{\varepsilon}_T = \frac{1}{2} (\dot{\varepsilon})_{\sigma_i = pos} + \frac{1}{2} (\dot{\varepsilon})_{\sigma_i = neg} \tag{3}$$

We next assume that the material in the areas of positive and negative internal stress deform by the same power law creep mechanism. We use the normal creep rate-stress relationship proposed for power law creep$(16-18)$ and include the internal stress:

$$\dot{\varepsilon}_T = \frac{SD}{2b^2} \left(\frac{\sigma_a + \sigma_i}{E}\right)^n + \frac{SD}{2b^2} \left(\frac{\sigma_a - \sigma_i}{E}\right)^n \tag{4a}$$

here, $\sigma_a$ is the applied stress, $\sigma_i$ is the magnitude of the average internal stress, E is the dynamic unrelaxed average polycrystalline modulus, D is the lattice diffusion coefficient, b is the Burgers vector, S is a dimensionless material constant, and n is the stress exponent. When $\sigma_i > \sigma_a$, the second term must contribute negatively to $\dot{\varepsilon}_T$, hence we rewrite equation (4a) to reflect this condition as follows:

$$\dot{\varepsilon}_T = \frac{SD}{2b^2} \left(\frac{\sigma_a + \sigma_i}{E}\right)^n + \frac{SD}{2b^2} \frac{(\sigma_a - \sigma_i)}{|\sigma_a - \sigma_i|} \left[\left|\frac{\sigma_a - \sigma_i}{E}\right|^n\right] \tag{4b}$$

The experimentally observed stress component, $n_{obs}$, can be expressed in terms of a diffusion-compensated creep rate and a modulus-compensated stress[16]:

$$n_{obs} = \frac{d \ln(\frac{\dot{\varepsilon}_T}{D})}{d \ln(\frac{\sigma_a}{E})} = \frac{d(\frac{\dot{\varepsilon}_T}{D})}{d(\frac{\sigma_a}{E})} \cdot \frac{(\frac{\sigma_a}{E})}{\frac{\dot{\varepsilon}_T}{D}} \tag{5}$$

Differentiating equation (4b), and realizing that the terms $\frac{\sigma_i}{E}$ and $\frac{(\sigma_a-\sigma_i)}{|\sigma_a-\sigma_i|}$ are constants, one obtains after substituting into equation (5):

$$n_{obs} = \frac{n\frac{\sigma_a}{E}\{(\frac{\sigma_a+\sigma_i}{E})^{n-1} + \frac{(\sigma_a-\sigma_i)}{|\sigma_a-\sigma_i|}[|\frac{\sigma_a-\sigma_i}{E}|^{n-1}]\}}{(\frac{\sigma_a+\sigma_i}{E})^n + \frac{(\sigma_a-\sigma_i)}{|\sigma_a-\sigma_i|}[|\frac{\sigma_a-\sigma_i}{E}|^n]} \tag{6}$$

The value of n in the above equation is determined by the mechanism actually controlling the plastic flow of the sample. The value of n is independent of any internal stress state. In the case of pure, polycrystalline uranium, this is most likely dislocation climb plus glide (slip creep), for which $n \approx 5$. Furthermore, $m_{obs} = \frac{1}{n_{obs}}$. Thus the strain rate sensitivity exponent experimentally measured during thermal cycling of pure, non-textured polycrystalline uranium can be expressed as:

$$m_{obs} = \frac{1}{n_{obs}} = \frac{(\frac{\sigma_a+\sigma_i}{E})^5 + \frac{(\sigma_a-\sigma_i)}{|\sigma_a-\sigma_i|}[|\frac{\sigma_a-\sigma_i}{E}|^5]}{\frac{5\sigma_a}{E}\{(\frac{\sigma_a+\sigma_i}{E})^4 + \frac{(\sigma_a-\sigma_i)}{|\sigma_a-\sigma_i|}[|\frac{\sigma_a-\sigma_i}{E}|^4]\}} \tag{7}$$

We now consider the following cases:
A) When $\sigma_a \gg \sigma_i$, $m_{obs} \approx 0.2$. Accordingly, during thermal cycling the macroscopically observed strain rate sensitivity exponent is not that much different from the microscopic (isothermal) strain rate sensitivity exponent.
B) When $\sigma_a \approx \sigma_i$, $m_{obs} \approx 0.4$. Accordingly, at sufficiently low applied stress so that the internal stress is about equal to the applied stress, the macroscopically observed strain rate sensitivity exponent is already close to the value (0.5) normally associated with fine structure superplastic behavior.
C) When $\sigma_i \approx 4\sigma_a$, $m_{obs} \approx 1$. Accordingly, at very low applied stress when the magnitude of internal stress is significantly larger than the applied stress, the macroscopically observed

strain rate sensitivity exponent approaches the ideal value for superplastic deformation. Recall from the section on the applied mechanics aspects of superplasticity that when $m_{obs} = 1$, the material is behaving in a Newtonian viscous manner, so that necking is suppressed and the material can be stretched indefinitely. Thus the concept of an internal stress as in [equation (4b)] can be used to explain qualitatively the superplastic behavior of thermally cycled pure, non-textured, polycrystalline alpha uranium.

Equation (4b) can be used to predict quantitatively the response of alpha uranium to thermal cycling. The terms n and $S/b^2$ in equation (4b) are the same for isothermal or thermal cycling conditions. From the isothermal data in Figure 6, n and $S/b^2$ are calculated to be 5.88 and $5.27 \times 10^{29}$, respectively. The parameter $\frac{\sigma_i}{E}$ cannot be measured experimentally, but a reasonable estimate can be made by determining what value of $\frac{\sigma_i}{E}$ will permit a theoretical prediction to intersect the experimental data in Figure 6 at one point. In this manner, it can be determined how well equation (4b) follows the general trend of the data. When $\frac{\sigma_i}{E} = 1.8 \times 10^{-4}$, equation (4b) exactly predicts the experimental value of $\frac{\dot{\varepsilon}_T}{D}$ at $\frac{\sigma_a}{E} = 10^{-5}$. Thus, for the experimental conditions described in Figure 6, $\frac{\sigma_i}{E}$ was estimated to be $1.8 \times 10^{-4}$. A theoretical prediction of the response of non-textured, polycrystalline alpha uranium to thermal cycling using equation (4b) can now be made. This prediction is shown in Figure 6 as a dashed line. The predicted behavior follows the experimental data reasonably well, suggesting that the model for internal stress superplasticity represented by equations (4) and (6) is a reasonable one.

Another example of internal stress superplasticity can be found amongst crystalline materials that undergo phase changes during creep flow. Steels will exhibit high strain rate sensitivity when creep tested at low stress during continuous thermal cycling through a phase transformation. We illustrate examples of several steels, taken from the work of Oeschlagel and Weiss[25], in Figure 8. As can be seen, elongations exceeding 400% are readily achieved. This type of internal stress superplasticity has been studied by a number of investigators. Historically, Sauveur[26] in 1929 may have been the first to discover this type of plastic flow when he twisted samples of iron in a temperature gradient. He noted that such samples were super weak in those regions where the alpha and gamma phases coexisted (i.e. where the alpha-gamma phase transformation was likely occurring during deformation). More recently, Rathenau and de Jong[27] showed that the transformation-induced strain under stress was directly proportional to the volume change during transformation (a measure of the internal stress) and to the applied stress (a linear dependence at low stresses), but inversely proportional to the inherent strength (hardness) of the matrix.

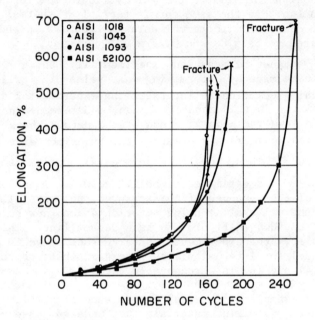

Figure 8. The above graph illustrates internal stress superplasticity, through multiple phase transformations, in several steels. As can be seen, the influence of cycling across the phase transformation temperature yields high elongations to fracture for several steels. The applied stress was 2500 psi. (Data from Oelschlagel and Weiss[25]).

# FUNDAMENTALS OF SUPERPLASTICITY

Their suggestion is given credence by the correlation[27-30] shown in Figure 9.

Application of internal stress superplasticity to industrial forming operations has been attempted by Weiss and his colleagues[31] and by Johnson[32]. For example, the warm drawing of tubes and threaded rods by dieless forming techniques have been attempted. The basic concept was to apply a small stress to the rod or tube while a heating device was moved at a specific speed along the work piece. Good results were achieved in the laboratory but apparently were never developed to a sufficiently controllable state for production applications. Also, the economics of such processing operations may have been prohibitive in relation to normal drawing and machining operations.

(B) <u>Fine Structure Superplasticity</u>

Fine structure superplasticity refers to the high elongations obtainable in metallic alloys that have ultrafine grain sizes. Typically the grain size is less than five microns (5μm). The mode of plastic deformation is believed to be primarily that of grain boundary shearing and grain boundary migration; evidence of slip creep processes appear to be negligible. An example of a fine grained superplastically deformed ultrahigh carbon steel[33] is shown in Figure 10. As can be seen, an elongation of about 1200% was achieved before failure occurred. The transmission electron micrograph shown below the sample reveals equiaxed ferrite grains about two microns in diameter in the deformed area of the sample. The micrograph also reveals fine particles of iron carbide (cementite, $Fe_3C$) well distributed along the ferrite grain boundaries. The cementite particles inhibit the growth of the ferrite grains during superplastic deformation.

Fine structure superplasticity has been known as long as internal stress superplasticity. Rosenhain et al.[34] observed such behavior in materials in 1920, and Pearson[35] documented and publicized the extraordinary ductility of lead alloys (Pb-Sn and Pb-Bi) in 1934. Bochvar[36] coined the term "sverhplastichnost" (ultrahigh plasticity) in his studies on superplastic Al-Zn alloys in 1945. It was not until 1962, when Underwood[37] publicized the works of Soviet scientists, that interest in the United States took hold. Work by Backofen[38], Hayden[39], Weiss[40] and others[40-47] was soon much in evidence. Several reviews have been published on fine structure superplasticity[48-57]. It is now quite well established that fine structure superplastic metallic alloys exhibit m values in the order of 0.5 (n = 2). Theories have been developed to explain this stress dependence of superplastic flow[3,38,58,59]. It is generally accepted that superplastic flow is associated principally with grain boundary shearing processes. An example of a logarithm stress versus logarithm strain rate curve for a nickel-base

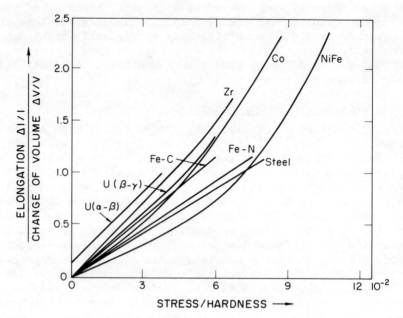

Figure 9. The above graph illustrates that the strain obtained upon transformation can be written as

$$(\varepsilon_{transf.}) = K \left(\frac{\Delta V}{V}\right) \frac{\sigma}{\text{Hardness}}$$

Thus the transformation strain is seen to be approximately a linear function of the applied stress. Furthermore, it is seen to be proportional to the volume change upon transformation (a measure of internal stress) and inversely proportional to the hardness (a measure of creep strength without internal stress). (Data from references 27-30)

# FUNDAMENTALS OF SUPERPLASTICITY

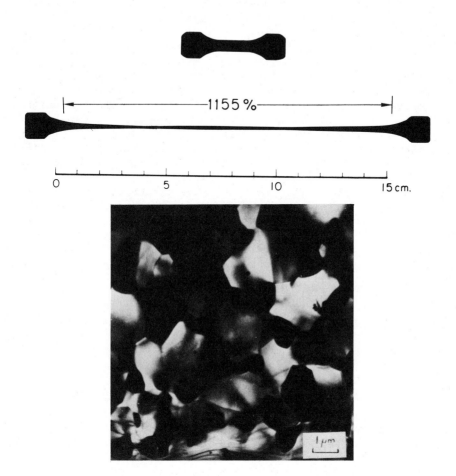

Figure 10. Superplastic ultrahigh carbon steel (1.6%C) can be deformed to elongations exceeding 1000% at 650°C as shown in the above sample ($\dot{\varepsilon}$ = 1% per minute). This high ductility is attributed to the presence of a fine ferrite grain size (~2µm) stabilized by fine cementite particles, as shown in the accompanying TEM photomicrograph.

superplastic alloy is shown in Figure 11. The figure illustrates the wide range of strain rate over which m = 0.5 (n = 2), and a second range at high strain rate where m decreases to a low value (typically m ≃ 0.15). In this high strain rate range plastic flow by diffusion controlled dislocation creep is known to prevail. Some researchers, however, consider that superplastic materials exhibit three discrete regions, as shown in Figure 12. Regions 2 and 3 are characterized by unique deformation processes. Region 2 represents the contribution of grain boundary shearing and migration as one unique process (m = 0.5, n = 2). Region 3 represents a slip creep process (i.e., diffusion-controlled dislocation creep), where m ≃ 0.12 to 0.20 (n ≃ 5-8), as another unique process. Region 1, the very low strain rate region, exhibits low m values and presumably non-superplastic behavior. This region is attributed by some investigators to a return of slip-creep like behavior involving two processes that are not independent, but instead, interlinked (i.e. sequential processes)[60,61]. Other researchers attribute this region to the presence of a threshold stress associated with grain boundary surface tension effects[62]. Another interpretation is that this region is to be associated with grain growth[63]; this is reasonable since the low strain rates used in this range means that the time of creep testing is long and grain growth is more likely to occur. An increase in grain size means that the strength will increase and be above that extrapolated from region 2 (shown in Figure 12). Such grain growth, leading to different grain sizes at each strain rate, will result in low apparent m values in region 1. In subsequent discussions of fine structure superplasticity we will confine ourselves to regions 2 and 3. We believe that classification of region 1 awaits further experimental work.

(1) <u>Prerequisites for fine structure superplasticity</u>

The prerequisites for superplastic behavior of fine structure metallic alloys are fairly well defined. Some of these suggested requirements are well established, whereas others are still under some debate or lack extensive definitive evidence. We outline some prerequisites below and indicate where controversy still exists.

(a) The major requirement is that the grain size should be fine. Typically, grains should be 5μm or less if superplastic behavior is to be observed at reasonable strain rates ($\dot{\varepsilon} \approx 10^{-4}$ sec$^{-1}$). Oftentimes high elongations are obtained for solid solution alloys where the grain size is fairly coarse (~20μm). The elongations observed (~200-300%) can be attributed to the presence of a solute-dislocation drag mechanism where the stress exponent is three[59,64,65]; the respectably high value of the strain rate sensitivity exponent (m = 0.33) can explain the high ductilities obtained. These highly ductile solid solution alloys should be not classified in the same category as fine structure superplastic alloys where m ≃ 0.5.

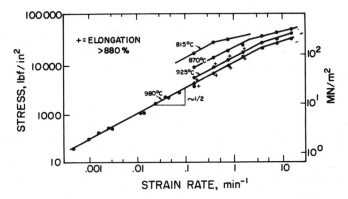

Figure 11. Flow stress-strain rate relationship (log-log plot) for a two phase nickel-iron-chromium alloy at elevated temperatures (39%Cr-10%Fe-1.75%Ti-1%Al-balance Ni). The strain rate sensitivity exponent is constant and high (m ≃ 0.5) over a wide range of strain rate. Elongations exceeding 880% (extent of the machine travel) were commonly observed. (Reference 39)

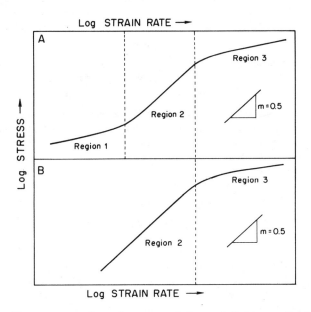

Figure 12. Many superplastic materials exhibit an S shaped behavior when plotted as logarithm of the flow stress versus logarithm of the strain rate (shown as A in the diagram above). Each of the three regions (making up the S shaped curve) are often associated with specific mechanisms of plastic flow. Other materials exhibit only two regions, as shown in schematic diagram B; an example of this type of behavior in a nickel-chromium-iron alloy is shown in Figure 11.

(b) In order to maintain a fine grain size at the superplastic forming temperature a second phase is usually, but not always, required. The second phase inhibits the matrix grains from growing. Greater inhibition of grain growth usually occurs as the amount of second phase is increased.

(c) The second phase in two phase alloys should be similar in strength to, or weaker than, the matrix phase. If the second phase is very hard, cavitation may occur at the interphase boundary during superplastic flow, leading to premature failure[66]. Also, the second phase, if very hard, may inhibit grain boundary sliding and migration. Examples of two phase alloys where superplastic behavior has not been observed are the oxide dispersion-strengthened alloys such as $Cu-Al_2O_3$, T-D nickel and T-D nichrome.

(d) The temperature of superplastic behavior is generally in the range 0.4 to $0.7T_m$. Above $0.7T_m$, grain growth can often readily occur despite the second phase, leading to a loss of superplasticity. Below $0.4T_m$, slip creep generally predominates as a principal mode of plastic deformation and again superplasticity is lost.

(e) The strain rates for superplastic flow are generally in the order of $10^{-6}$ to $10^{-3}$ per second. As will be seen later, this can be made to vary appreciably depending on grain size stability, ease of slip creep and other factors.

(f) The grain boundaries should be high angle (disordered) boundaries. An example of this important factor is given in Figure 13 for the case of a eutectoid composition steel[14].

(g) The grain boundaries should be mobile. Grain boundary mobility permits elimination of stress concentrations at triple points developed during grain boundary shearing. Marya and Wyon[67] made aluminum-gallium alloys superplastic at room temperature by increasing the ease of boundary shearing and migration through introduction of a low melting solid solution region of Al-Ga in the vicinity of the grain boundary.

(h) The grain boundaries should not be prone to ready tensile separation. This factor may be a major reason why fine grain size polycrystalline ceramics are not superplastic even when m is nearly unity. The ease of tensile separation at grain boundaries for such materials may be due to the high grain boundary energy inherent in such materials[68]. Another reason why polycrystalline ceramics are not superplastic is perhaps due to the inability of grain boundaries to migrate readily. Further studies on the behavior of ceramic materials at elevated temperature are needed to clarify some of these issues.

(i) The morphology of grains are such that they should be equiaxed. Elongated grains cannot exhibit extensive grain boundary shearing.

# FUNDAMENTALS OF SUPERPLASTICITY

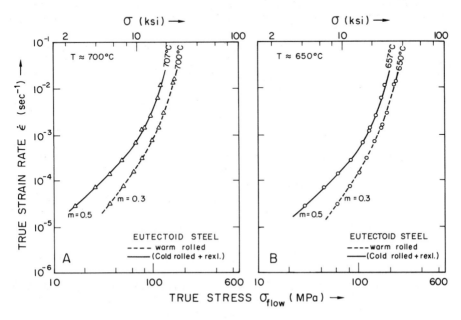

Figure 13. When a eutectoid composition steel is warm worked, a fine structure of ferrite grains and cementite particles is developed. The ferrite grain boundaries, however, consist of a mixture of high and low angle boundaries. Such a material is not superplastic; the m values are typically about 0.3 and only 100% elongation is obtained. If the steel is cold worked and recrystallized, high angle boundaries are obtained and the material is superplastic (m = 0.5 and elongations exceeding 400% are readily achieved). (Reference 14)

(2) <u>Phenomenological equation for fine structure superplasticity</u>

The factors influencing superplastic flow have been investigated extensively. Several theories of superplastic flow have been proposed and developed in a quantitative manner[3,38,58-60,62]. None of these theories, however, appear to have clearly resolved the collective contributions of the three major factors influencing superplasticity, namely: (1) the temperature dependence of superplastic flow, (2) the grain size effect, and (3) the stress dependence of superplastic flow. We have made an extensive study[69] on the phenomenology of superplastic flow in a large number of systems. This study has resulted in the following relation for superplastic flow:

$$\dot{\varepsilon}_{spf} = 2 \times 10^9 \frac{D^*_{eff}}{d^2} \left(\frac{\sigma}{E}\right)^2 \quad (8)$$

In this relation $\dot{\varepsilon}_{spf}$ is the superplastic flow rate (region 2 of Figure 12), $D^*_{eff}$ is the effective diffusion coefficient, d is the grain size during superplastic flow, $\sigma$ is the flow stress and E is the dynamic unrelaxed average Young's modulus. The effective diffusion coefficient is given by

$$D^*_{eff} = D_L f_L + c\, D_{gb}\, f_{gb} \quad (9)$$

where $D_L$ is the lattice diffusion coefficient, $D_{gb}$ is the grain boundary diffusion coefficient, $f_L$ is the fraction of atoms associated with lattice diffusion (about unity), $f_{gb}$ is the fraction of atoms associated with grain boundary diffusion and c is a constant equal to 0.01. The term $f_{gb}$ is generally given as $\frac{\pi w}{d}$, where w is the grain boundary width (typically considered equal to 2b, where b is the Burgers vector). Equation (9) is unusual in that the factor c is equal to 0.01, a constant needed to predict accurately the temperature dependence of superplastic flow and to predict conditions where lattice diffusion or grain boundary diffusion will be the rate controlling process. The factor c is normally considered to be unity, as in the case of Ashby and Verrall's[62] unification of Nabarro-Herring and Coble creep into one theory involving stress-directed diffusional creep.

Equations (8) and (9) can be rewritten explicitly for the case where grain boundary diffusion is the rate controlling process and for the case where lattice diffusion is rate controlling. Thus when $.01\, D_{gb}\, f_{gb} \gg D_L f_L$, then $D^*_{eff}$ becomes equal to $2 \times 10^{-2} D_{gb} \frac{\pi b}{d}$. Substituting this expression for $D^*_{eff}$ in equation (8) yields

$$\dot{\varepsilon}_{spf} \simeq 10^8 \frac{D_{gb}\, b}{d^3} \left(\frac{\sigma}{E}\right)^2 \quad \text{(grain boundary diffusion controlled)} \quad (10a)$$

In the case when lattice diffusion controls the superplastic

# FUNDAMENTALS OF SUPERPLASTICITY

flow process, $D_L >> 2 \times 10^{-2} \frac{D_{gb} \pi b}{d}$, and therefore $D^*_{eff} = D_L$. Substituting the expression for $D^*_{eff}$ in equation (8) yields

$$\dot{\varepsilon}_{spf} = 2 \times 10^9 \frac{D_L}{d^2} \left(\frac{\sigma}{E}\right)^2 \tag{10b}$$

We illustrate the general validity of the phenomenological equation for superplastic flow with two examples. The first example is shown in Figure 14. This figure illustrates the excellent prediction of the temperature dependence of a 35Cr-39Ni-26Fe alloy[70] by means of equation (9). Whereas the normal $D_{eff}$ (where c of equation (9) equals unity) would predict a transition from $D_{gb}$ to $D_L$ at near 1600°C (200°C above the melting temperature), the relation for $D^*_{eff}$ given in Equation (9) predicts a transition temperature of 900°C. This is exactly at the break in the slope of the logarithm creep versus reciprocal temperature curve for the 35Cr-39N9-26Fe alloy. The second example relates to the grain size prediction given in equations (10a) and (10b). Equation (10a) predicts that, when grain boundary diffusion is controlling superplastic flow, the grain size exponent should be 3. Equation (10b) predicts that the grain size exponent should be 2 when lattice diffusion is controlling fine structure superplastic flow. These predictions are consistent with the grain size exponents reported in the literature (Table 1).

### (3) Optimizing the rate of superplastic flow in fine structure materials

Earlier in this paper we illustrated in a schematic sense (Figure 12) the range of strain rate and stress where superplastic flow dominates the deformation process (region 2). Above a certain specific strain rate, superplastic flow is no longer the dominant process and another mode of deformation becomes important, namely, diffusion controlled dislocation creep (slip creep). The maximum creep rate for superplastic flow is typically in the order of $10^{-4}$ to $10^{-3}$ s$^{-1}$, a rate considerably slower than those used in commercial forming operations: the strain rate in commercial forming operations is commonly about 1 s$^{-1}$. From a technological viewpoint it would be highly desirable to increase the maximum strain rate for superplastic flow. Since the factors influencing superplastic creep differ from those influencing slip creep, the approach to reach this goal is straight forward. One must select variables that will enhance superplastic flow but make slip creep more difficult. We illustrate this schematically in Figure 15. In this graph, the logarithm of the strain rate is plotted against the logarithm of the flow stress. In such a plot, the slope is equal to the stress exponent, n, ($n = \frac{1}{m}$). The two separate processes contributing to region 2 and region 3 are represented as straight lines, and the point of intersection (marked as 1) represents the maximum strain

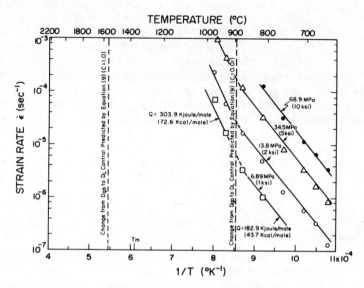

Figure 14. Arrhenius plot of Hayden, Floreen and Goodell data for a 35 Cr-39 Ni-26 Fe alloy showing the good agreement between the temperature where the change from grain boundary diffusion-controlled to lattice diffusion-controlled superplastic flow occurs and that predicted by equation (9) with c = 0.01. The predicted temperature for this change as given by $D_{eff}$ with c = 1 is also given.

## Table 1

Grain Size Exponent, p, as Reported in the Literature Compared to that Predicted by Phenomenological Equations (10a) and (10b) for Fine Structure Superplastic Flow

| Material | Diffusion Control Type | Reported Grain Size Exponent (p) | Predicted Grain Size Exponent (p) | Reference |
|---|---|---|---|---|
| 67 Al-33 Cu | $D_L$ | 2.1-2.4 | 2 | 44 |
| 67 Al-33 Cu | $D_L$ | 2 | 2 | 71 |
| 93 Cu-7 P | $D_L$ | 1.98-2.04 | 2 | 72 |
| IN 744 | $D_L$ | 2 | 2 | 70 |
| 35 Cr-39 Ni-26 Fe | $D_L$ | 2 | 2 | 73 |
| 35 Cr-39 Ni-26 Fe | $D_{gb}$ | 3 | 3 | 70 |
| 95 Pb-5 Cd | $D_{gb}$ | 2.6 | 3 | 74 |
| 62 Sn-38 Pb | $D_{gb}$ | 2.3 | 3 | 75 |
| 62 Sn-38 Pb | $D_{gb}$ | 3 | 3 | 42 |
| 95 Sn-5 Bi | $D_{gb}$ | 3 | 3 | 40 |
| 78 Zn-22 Al | Mixed | 2 | - | 3 |
| 78 Zn-22 Al | Mixed | 2 | - | 76 |
| 78 Zn-22 Al | Mixed | 4.5 | - | 77 |

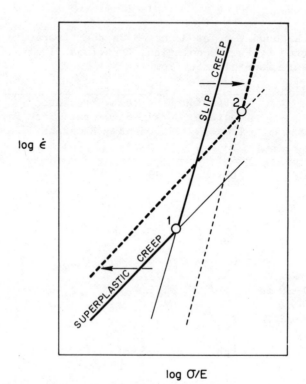

log σ/E

Figure 15. The solid line depicts the creep behavior of a fine structure superplastic material as a function of stress at a given structural state. Point 1 on the curve represents the maximum strain rate where superplastic behavior can be expected. If the structural state is changed such that superplastic flow is enhanced and slip creep is made more difficult, as shown by the dashed lines, then the maximum strain rate for superplastic flow is increased (point 2).

# FUNDAMENTALS OF SUPERPLASTICITY

rate for superplastic flow for a given microstructural condition. Dashed lines are given to indicate what a new set of microstructural conditions can do if they lead to increasing the superplastic flow rate and to decreasing the slip creep rate. Such a change then leads to an increase in the maximum rate for superplastic flow, as given by symbol 2 in Figure 15.

The major microstructural feature influencing superplastic flow is the grain size. When grain boundary diffusion is rate controlling the superplastic flow rate is inversely proportional to the cube of the grain size [Equation (10a)]. An example of the predicted enhancement of superplastic flow by control of grain size is shown for an ultrahigh carbon steel[33] in Figure 16. If the grain size is decreased from 2µm to 0.4µm, it is predicted that the maximum strain rate for superplastic flow is enhanced by a factor of about a thousand. Maintaining a grain size of less than one micron during superplastic flow is not a trivial problem and represents an important area of research for the development of commercially useful superplastic alloys.

Diffusion controlled dislocation creep (slip creep) is influenced by a number of microstructural variables[16-18]. Slip creep can be decreased by changing the crystal structure[16] (b.c.c., h.c.p., f.c.c., diamond cubic, ordered structures, etc.), by texture[73], by substructure strengthening[79], and by decreasing the stacking fault energy[80]. We illustrate in Figure 17 the possible enhancement of superplastic behavior in nickel as influenced by stacking fault energy. The creep rate of nickel in the slip creep region is decreased by solid solution alloying additions of tungsten (81) to lower the stacking fault energy. The predicted enhancement in the maximum superplastic strain rate of a 2µm grain size nickel alloy is by a factor of twenty with the addition of six atomic percent tungsten to nickel.

## (C) Superplastic solid state bonding

Fine structure superplastic metallic alloys are characterized by the presence of fine grains. Grain boundaries are regions of high atom mobility. Typically, the diffusion coefficient in grain boundaries is about $10^{-9}$ cm$^2$/sec at $0.5T_m$; this is about $10^6$ times greater than in the lattice of a b.c.c. material and $10^9$ greater than in the lattice of an f.c.c. or h.c.p. material. Such high atom mobility would suggest that solid state bonding of superplastic materials may be readily achieved at warm temperatures and low pressures. Indeed, Caligiuri[82] has shown that superplastic ultrahigh carbon steel powders densify more readily than non-superplastic pure iron powders at 650°C ($0.51T_m$) when a pressure is applied to cause the material to creep at the same rate. An example is shown in Figure 18. The effect observed may, in part, be attributed to the availability of many high diffusivity paths at adjoining surfaces

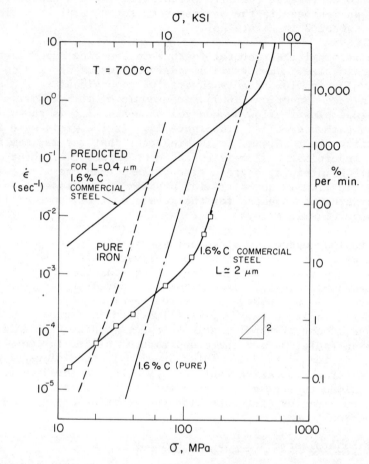

Figure 16. Grain size is the most important single factor influencing the range of strain rate over which superplastic flow is observed since $\dot{\varepsilon}$ superplastic is proportional to $d^{-3}$ (equation (10a), whereas $\dot{\varepsilon}$ slip creep is not a function of grain size. The above graph illustrates the strain rate-flow stress relationship for a 1.6%C steel at 700°C. For a grain size of d = 2μm, superplastic flow is expected up to a strain rate of 10% per minute. However, if the grain size is decreased to 0.4μm, superplastic flow can be expected to occur up to strain rates as high as 5000% per minute. (Reference 33).

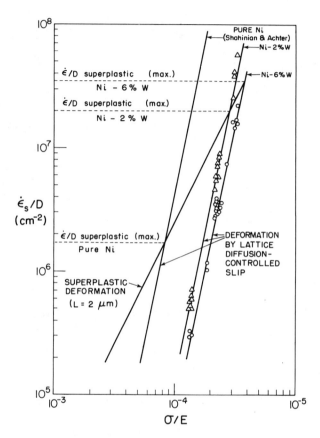

Figure 17. Tungsten enhances the creep resistance of nickel in the slip creep region by decreasing the stacking fault energy. On the other hand, $\dot{\varepsilon}$ superplastic is unaffected by stacking fault energy. Thus, the extent of superplastic flow may be broadened by solute additions which decrease the stacking fault energy of the matrix. (Ni - W data from reference 81).

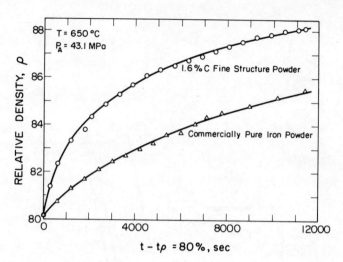

Figure 18. Fine structure superplastic powders (1.6%C steel) densify more readily than commercially pure iron powders when pressure sintered at a stress where the creep rates of the two materials are identical. The temperature of densification was 650°C and the applied pressure was 43.1 MPa (6.2 ksi). (Data from Reference 82).

of the powders. Caligiuri showed that the densification rate of powders is a function of the creep rate according to the following relation

$$\dot{\rho} \simeq 60 \left[ \frac{\rho_t}{\rho} - 1 \right]^n \dot{\varepsilon}_{ss} \qquad (11)$$

where $\dot{\rho}$ is the relative densification rate [i.e., $d(\frac{\rho}{\rho_0})/dt$], $\rho$ is the density, $\rho_t$ is the theoretical density and $\dot{\varepsilon}_{ss}$ is the steady state creep rate at the applied stress. The important point regarding equation (11) is that the densification rate is proportional to the creep rate and also to the creep mechanism involved in the plastic flow process through the term $(\frac{\rho_t}{\rho} - 1)^n$, where n is the stress exponent for creep. The relative ease with which powders of superplastic materials are pressure sintered suggests that such materials can be readily bonded when pressed as plates[82]. An example of the excellent bond obtained between two superplastic ultrahigh carbon steel plates when pressed at 650°C and a stress of 69 MPa (10 ksi) is shown in Figure 19. The pressing was performed in air. The bond is excellent. Bend tests on the bonded plates revealed that separation will not occur at the interface of the plates. Such results suggest many new applications in the area of warm temperature solid state bonding or welding technology.

# FUNDAMENTALS OF SUPERPLASTICITY

## Summary

Superplastic metallic alloys have mechanical characteristics which can make them ideal for both the fabricator and the user of materials. The fabricator has a material which exhibits extraordinary tensile elongation ($\geq$ 400% elongation) and low resistance to plastic flow at warm temperatures (0.4 to $0.7T_m$). Such materials can be readily formed into near-net-shape final products, thereby avoiding expensive machining, joining and fastening operations (Figures 1 and 2). The user in turn has a material which can have high strength and ductility. In many cases, superplastic alloys are amenable to post-shaping heat treatment for enhancement of end use properties. Superplastic materials may find use for joining applications through solid state diffusion bonding or welding (Figure 19).

The basis of superplasticity is the high strain rate sensitivity characteristic of all superplastic metallic alloys. Applied mechanics permits a calculation of the percentage elongation to fracture as a function of the strain rate sensitivity exponent, m, in the equation $\sigma = K\dot{\varepsilon}^m$ ($\sigma$ is the flow stress, $\dot{\varepsilon}$ is the strain rate and K is a material constant). Good correlation is obtained between the predicted behavior and that observed experimentally (Figure 5).

Two types of superplasticity have been noted. One is internal stress superplasticity. This type of superplasticity is associated with the development of internal stress during plastic flow. The volume change during phase transformation by thermal cycling polymorphic metals is one example of internal stress superplasticity (Figure 8). Another example is the internal stress that is generated by thermally cycling polycrystalline metals that have a large anisotropy in their thermal expansion coefficient (Figure 7). These internal stresses assist plastic flow and lead to high strain rate sensitivity, m values being as high as unity. The second type of superplasticity is associated with materials that have a fine grain size, typically less than 5µm. Grain boundary shearing and migration appears to dominate the microstructural features accompanying plastic flow in fine structure superplastic materials. Phenomenological studies on such materials reveal that the strain rate sensitivity exponent, m, is typically about equal to 0.5 (Figure 11). The superplastic flow rate is controlled either by lattice diffusion or by grain boundary diffusion and can be described by means of an effective diffusion coefficient (Equation 9). When lattice diffusion is rate controlling, the creep rate proportional to $d^{-2}$ where d is the grain size, and when grain boundary diffusion is rate controlling, the creep rate is proportional to $d^{-3}$ (Equations (10a) and (10b) and Table 1).

Fine structure superplastic materials lose their superplastic effect as the strain rate is increased. This is due to the in-

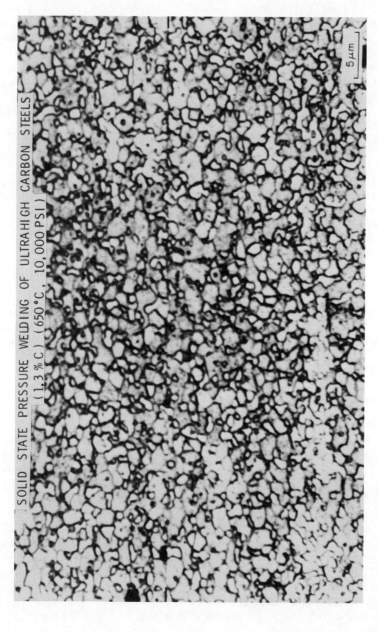

Figure 19. The above photomicrograph illustrates that superplastic ultrahigh carbon steel plates (1.3%C) can be readily solid state diffusion bonded in air at relatively low temperature ($0.51T_m$) and low pressure (69 MPa). A line of demarcation can be seen at the bond interface because the cementite particles and ferrite grains are flattened at the interface, leading to many flat ferrite grain and interphase boundaries. (Reference 32).

# FUNDAMENTALS OF SUPERPLASTICITY

creasing contribution of slip creep to the deformation process with increasing strain rate (Figure 15). The maximum strain rate for superplastic flow can be increased by enhancing superplastic flow through grain size refinement (Figure 16), or by inhibiting slip creep through control of stacking fault energy (Figure 17), texture, crystal structure and dislocation substructure within the grains.

## ACKNOWLEDGEMENT

The authors would like to acknowledge support from the Office of Naval Research and the Defense Advanced Research Projects Agency. They wish to thank especially Drs. B. MacDonald, A. Bement, and E. Van Reuth for their guidance and encouragement. In addition, the authors acknowledge valuable technical contributions for Dr. Jeffrey Wadsworth.

## References

1. J. B. Moore and R. L. Athey, U. S. Patent 3,519,503.

2. C. Howard Hamilton and George W. Stacher, Metal Progress, March 1976, 24-37.

3. A. Ball and M. M. Hutchison, Met. Science J., 3, (1969), 1.

4. D. A. Woodford, Trans. ASM, 62, (1969), 291.

5. D. H. Avery and W. Backofen, Trans. ASM, 58, (1965), 551.

6. A. W. Mullendore, and N. J. Grant, Trans. AIME, 200, (1954), 973.

7. J. B. Conway and P. N. Flagella, Creep-Rupture Data for the Refractory Metals to High Temperatures, Nuclear Systems Programs Technical Report GEMP-685 [R-69-NSP-9], General Electric Co.

8. W. Green, Trans. AIME, 215, (1959), 1057.

9. B. A. Wilcox and A. H. Clauer, Trans. AIME, 236, (1966), 570.

10. C. Rossard, Rev. Met., 63, (1966), 225.

11. W. B. Morrison, Trans. ASM, 61, (1968), 423.

12. D. H. Avery and J. M. Stuart 1968, in Surfaces and Interfaces II, Physical and Mechanical Properties, ed. J. J. Burke, N. L. Reed and V. Weiss, Syracuse University Press, Syracuse, New York, p. 371.

13. M. A. Burke, and W. D. Nix, Acta Met., 23, (1975), 793.

14. E. S. Kayali, Ph.D. Dissertation, Department of Materials Science and Engineering, Stanford University, 1976.

15. O. D. Sherby et al., Scripta Met., 9, (1975), 569.

16. O. D. Sherby and P. M. Burke, Progress in Materials Science, B. Chalmers and W. Hume Rothery, Editors, Pergamon Press, Oxford, 13, (1967) 325.

17. A. K. Mukherjee, J. E. Bird and J. E. Dorn, Trans. ASM, 62, (1969) 155.

18. J. Weertman, Trans. ASM, 61, (1968) 681.

19. L. T. Lloyd and C. S. Barrett, J. Nucl. Mater., 18, (1966) 55.

20. R. C. Lobb, E. C. Sykes and R. H. Johnson, Met. Sci. J., 6, (1972), 33.

21. S. L. Robinson, O. D. Sherby and P. E. Armstrong, J. Nucl. Mater., 46, (1972), 293.

22. O. D. Sherby, D. L. Bly and D. H. Wood, in Physical Metallurgy of Uranium Alloys, J. J. Burke, D. A. Colling, A. E. Gorum and J. Greenspan, Editors, Brook Hill Publishing Company, 1976, 311.

23. Y. Adda, and A. Kirianenko, J. Nuc. Mat., 6, (1962), 130.

24. P. E. Armstrong, D. T. Eash, and J. E. Hockett, J. Nuc. Mat., 45, (1972) 211.

25. D. Oelschlagel, and V. Weiss, Trans ASM, 59, (1966), 143.

26. A. Sauveur, Iron Age, 113, (1924), 581.

27. M. de Jong and G. W. Rathenau, Acta Met., 9, (1961), 714,

28. R. H. Johnson and G. W. Greenwood, Nature, 195, (1962), 138.

29. G. W. Greenwood and R. H. Johnson, Proc. Roy. Soc., A283, (1965) 403.

30. W. Chubb, Trans. AIME, 203, (1955), 189.

31. V. Weiss and R. Kot, Wire Journal, September 1969.

32. R. H. Johnson, Design Engineering, March 1969.

33. B. Walser and O. D. Sherby, Met. Trans. AIME, in press, 1978.

34. W. Rosenhain, J. L. Haughton and K. E. Bingham, Inst. of Metals, 23, (1920), 261.

35. C. E. Pearson, J. Inst. Metals, 54, (1934), 111.

36. A. A. Bochvar and Z. A. Seiderskaya, Izvestia Acad, Nauk, USSR, OTN, 9, (1945) 821.

37. E. E. Underwood, J. Metals, 914, (1962), 919.

38. W. A. Backofen, I. R. Turner and D. H. Avery, Trans. ASM, 57, (1964), 980.

39. G. W. Hayden, R. C. Gibson, H. F. Merrick and J. H. Brophy, Trans. ASM, 60, (1967), 3.

40. W. Weiss and R. Kot, Proc. Int. Conf. Manufacturing Tech. (Am. Soc. Tool and Manuf. Engr.) 1967, 1031.

41. T. H. Alden, Acta Met., 15, (1967) 469.

42. C. M. Packer and O. D. Sherby, Trans. ASM, 60, (1967), 21.

43. D. L. Holt, Trans. AIME, 242, (1968), 25.

44. D. L. Holt and W. A. Backofen, Trans. ASM, 59, (1966), 755.

45. P. Chaudhari, Acta Met., 15, (1967), 1777.

46. H. E. Cline and T. H. Alden, Trans. AIME, 242, (1968), 25.

47. D. Lee and W. A. Backofen, Trans. AIME, 239, (1967), 1034.

48. O. D. Sherby, Science Journal, 5, (1969), 75.

49. J. J. Burke and V. Weiss, 1970 Ultrafine Grain Metals, Syracuse University Press, Syracuse, New York.

50. R. H. Johnson, Met. Rev., 15, (1970), 115.

51. G. J. Davies, J. W. Edington, C. P. Cutler and K. A. Padmanabhan, J. Mater. Sci., 5, (1970), 1091.

52. R. B. Nicholson, in Electron Microscopy and the Structure of Materials, ed. G. Thomas, R. M. Fulrath, and R. M. Fisher, University of California Press, Berkeley, 1972, 689.

53. A. K. Mukherjee, in Treatise on Materials Science and Technology, ed. R. J. Arsenault, Academic Press, New York, 6, 1975, 163.

54. T. H. Alden, in *Fundamental Aspects of Structural Alloy Design*, eds. R. I. Jaffee and B. A. Wilcox, Plenum Press, 1977, 411.

55. M. H. Shorshorrov, A. S. Tichonov, C. I. Bulat, K. P. Gurov, N. I. Nadirashvili, and V. Z. Antipov, *Superplasticity in Metallic Materials*, Uzdatelsvo, "Nauk", Moscow, 1973.

56. John Gittus, Creep, Viscoelasticity and Creep Fracture in Solids (Chapter II), Applied Science Publishers Ltd., London 1975, 509.

57. J. W. Edington, K. N. Melton and C. P. Cutler, Progress in Materials Sci., B. Chalmers, J. W. Christian and T. B. Massalski, Eds., 21, 1976, 61.

58. T. G. Langdon, Phil. Mag., 22, (1970), 689.

59. A. K. Mukherjee, Mat. Sci. & Engr., 8, (1971), 83.

60. R. C. Gifkins, Met. Trans., 7A, (1976), 1225.

61. T. G. Langdon and F. A. Mohammed, Jnl. Australian Inst. Metals, 22, (1977), 189.

62. M. F. Ashby and R. A. Verrall, Acta Met., 21, (1973), 149.

63. G. Rai and N. J. Grant, Met. Trans., 6A, (1975), 385.

64. J. R. Cahoon, Met. Sci. J., 9, (1975), 346.

65. R. C. Gifkins, J. Inst. Metals, 95, (1967), 373.

66. S. Sagot, P. Blenkinsup and D. M. R. Taplin, J. Inst. Metals, 100, (1972), 268.

67. S. K. Marya and G. Wyon, Fourth International Conf. on the Strength of Metals and Alloys, Nancy, France, 1, September 1976, 438.

68. J. J. Gilman, in *Mechanical Behavior of Crystalline Solids*, National Bureau of Standards Monograph 59, 1963, 79.

69. R. A. White, Ph.D. Dissertation, Department of Materials Science and Engineering, Stanford University, Stanford, Ca., March 1978.

70. P. D. Goodell, S. Floreen and H. W. Hayden, Met. Trans., 3, (1972), 833.

71. M. J. Stowell, J. L. Robertson and B. M. Watts, Mat. Sci, J., 3, (1969), 41.

72. C. Herriot, M. Suery and B. Baudelet, Scripta Met., 6, (1972), 657.

73. H. W. Hayden and J. H. Brophy, Trans. ASM, 61, (1968), 542.

74. T. H. Alden, Acta Met., Trans. ASM, 61, (1968), 559.

75. D. H. Avery and W. A. Backofen, Trans. ASM, 58 (1965), 551.

76. D. L. Holt, Trans. AIME, 242, (1968), 740.

77. T. H. Alden and H. W. Schadler, Trans. AIME, 242, (1968), 825.

78. G. R. Edwards, T. R. McNelley and O. D. Sherby, Scripta Met., 8, (1974), 475.

79. O. D. Sherby, R. H. Klundt and A. K. Miller, Met. Trans., 8A, (1977), 843.

80. C. R. Barrett and O. D. Sherby, Trans. TMS-AIME, 233, (1965), 1116.

81. W. R. Johnson, C. R. Barrett and W. D. Nix, Met. Trans., 3, (1972), 963.

82. R. D. Caligiuri, Ph.D. Dissertation, Department of Materials Science and Engineering, Stanford University, Stanford, Ca., August 1977.

ADVANCES IN THE HEAT TREATMENT OF STEELS

J. W. Morris, Jr., J. I. Kim and C. K. Syn

Department of Materials, Science and Mineral Engineering, and Materials and Molecular Research Division, Lawrence Berkeley Laboratory, University of California, Berkeley

ABSTRACT

A number of important recent advances in the processing of steels have resulted from the sophisticated uses of heat treatment to tailor the microstructure of the steels so that desirable properties are established. These new heat treatments often involve the tempering or annealing of the steel to accomplish a partial or complete reversion from martensite to austenite. The influence of these reversion heat treatments on the product microstructure and its properties may be systematically discussed in terms of the heat treating temperature in relation to the phase diagram. From this perspective, four characteristic heat treatments are defined: (1) normal tempering, (2) intercritical tempering, (3) intercritical annealing, and (4) austenite reversion. The reactions occurring during each of these treatments are described and the nature and properties of typical product microstructures discussed, with specific reference to new commercial or laboratory steels having useful and exceptional properties.

I. INTRODUCTION

A number of important advances in the processing of steels which have been made over the past twenty years have involved the sophisticated use of heat treatment to control the microstructure and consequently the properties of the steel. Some of the most useful of these new treatments involve variations on a simple elementary step: the alloy, which has been initially cooled or quenched to form a ferritic or martensitic starting structure, is

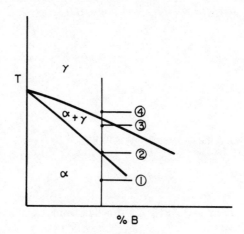

Figure 1: Schematic drawing showing four characteristics heat treatment points in relation to the equilibrium phase diagram.

reheated to an intermediate temperature to accomplish a partial or complete reversion to the high temperature austenite phase. The nature and extent of the reversion reaction and the details of the associated microstructural changes may, of course, depend in a rather complex way on the composition of the alloy and on the heating and cooling rates employed. If we restrict our attention to alloys of relatively low interstitial content, however, it is possible to distinguish a sequence of characteristic microstructural changes which occur as the reheating temperature is raised relative to the phase diagram of the steel. These characteristic microstructural changes form a useful framework for a discussion of the response of steels to reheating treatments.

The elementary types of response to reheating may be classified by considering a prototypic alloy, a portion of whose phase diagram is shown as Figure 1. Let the composition of the alloy be given by the vertical line in the figure, and assume that the alloy has initially been quenched to form a martensite structure. Let the alloy now be reheated to one of the four temperatures noted along the isocomposition line, held at the selected temperature for some time, and then recooled to room temperature, using heating and cooling rates which are sufficiently rapid that decomposition during the heating and cooling steps can be ignored. The four temperatures noted along the isocomposition line then define four characteristic reactions:

(1) Normal Tempering. If the steel is held at a temperature which is below the temperature at which the formation of the austenite phase begins, then the expected reactions are those associated with normal tempering, which are familiar from the technology of plain carbon and maraging steels. The steel may harden through the formation or reconfiguration of small precipitates, such as carbides, and may simultaneously recover through the equilibration of point defects and the relaxation of dislocations introduced during the martensite transformation. It is well known that the properties of such steels may be adjusted over a very wide range by controlling the tempering time and temperature.

(2) Intercritical Tempering. If the reheating temperature is adjusted to the point marked (2) in Figure 1, which is slightly within the two phase $\alpha + \gamma$ region, then the reactions occurring during tempering will also involve a precipitation of the high temperature austenite phase. If the tempering conditions are such that the austenite phase can equilibrate insofar as its solute content is concerned then the precipitated austenite will be rich in solute, relatively stable, and may hence be retained in the microstructure after the alloy is returned to room temperature. The final microstructure will then be a mixture of tempered martensite and precipitated austenite retained in the metastable austenite phase. The consequences of this intercritical tempering will be discussed in more detail below. This treatment is basic in the production of typical structural steels intended for use at cryogenic temperature, for example, commercial "9Ni" steel.

(3) Intercritical Annealing. If the reheating is to the temperature marked (3) in Figure 1, near the top of the two phase $\alpha + \gamma$ region, then a more extensive decomposition will occur. A much higher volume fraction of the austenite phase is anticipated and this austenite phase, according to the equilibrium diagram, will be much leaner in solute species. If the reheating temperature is high enough the precipitated austenite will be unstable with respect to martensite transformation on subsequent cooling to room temperature, and most of it will revert to the martensite phase. In this case the product microstructure will consist primarily of two components: a well-tempered martensite, or ferrite, which is relatively lean in solute, and a fresh martensite which is relatively high in solute and which is produced by the retransformation of the austenite precipitated during the heat treatment. The intercritical annealing reaction is fundamental to the processing of so-called "dual phase" steels, which are now under extensive development for automotive applications. Intercritical annealing is also used in combination with other heat treatments in the processing of specialty alloys intended for low temperature use.

(4) Austenite Reversion. If the martensitic steel is heated at a reasonably rapid rate to the temperature marked (4) in Figure 1, which lies well within the austenite stability field, then the alloy will undergo a complete reversion to the austenite phase. If the alloy is relatively rich in austenite stabilizing solutes, so that the reversion temperature need not be too high, then the reverse transformation to austenite will occur predominantly by a shear mechanism which reverses the martensitic transformation. If the steel in question is a martensitic steel, that is if its martensite transformation temperatures lie above room temperature, then the austenite transformation will be reversed on subsequent cooling. The final microstructure of the steel will then be predominantly martensitic, but may have a finer effective grain size and a much higher dislocation density because of the double-shear transformation. If the steel is an austenitic steel, in the sense that its martensite transformation temperatures lie below room temperature, and if it was originally transformed to martensite by cooling to some very low temperature, then the steel will remain austenitic after being returned to room temperature. Because of the double shear transformation, however, the austenitic steel will now contain a very high density of internal defects. The reversion reaction may hence be used to "transformation strengthen" an austenitic steel which would otherwise be relatively soft in the as-quenched condition.

Of these four elementary reactions, the normal tempering reaction has been familiar to steel metallurgists for a great many years. While increasing metallurgical sophistication has led to continued improvements in the use of tempering treatments and in the resulting properties of tempered or maraged steel, the elementary features of normal tempering are reasonably well known. On the other hand treatments which involve the use of retransformation to austenite as a method of achieving a desirable final microstructure are more recent in origin, and are only now beginning to be understood and efficiently exploited. In the following sections we shall discuss the three prototypic types of such treatments: intercritical tempering, intercritical annealing, and austenitic reversion, in more detail using specific examples of their utilization in the processing of research alloys and commercial steels. The illustrations will be drawn primarily from our own recent work and that of others at the Lawrence Berkeley Laboratory, not with the implication that this is the only, or even the best work in the area, but rather it is most familiar to us.

## II. INTERCRITICAL TEMPERING

If an initially martensitic alloy is reheated to the point marked (2) in Figure 1, just within the two-phase region, and held, then a solute-rich $\gamma$-phase will precipitate within the matrix of

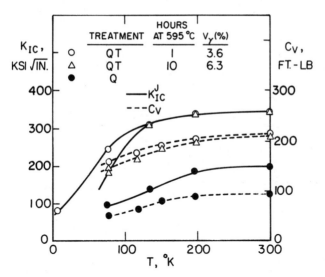

Figure 2: Fracture and impact toughness as a function of temperature for 9Ni steel in the as-quenched and intercritically tempered conditions.

the initial $\alpha'$. If the solute content of this precipitated austenite is sufficiently high a substantial fraction of it will be retained on subsequent cooling to create a two-phase mixture of tempered martensite and retained austenite at room temperature.

A typical use of intercritical tempering is in the processing of ferritic alloys intended for structural use at low temperature, including the established Fe-Ni[1-4] and recently developed Fe-Mn[5] grades. The simplest case is commercial "9Ni" steel, which is often used in a quench-and-tempered (QT) condition. Alloys which are leaner in nickel or which are intended for service under more extreme conditions are usually given a more elaborate treatment before the tempering step.

The intercritical tempering treatment is used to improve low-temperature toughness, principally by lowering the ductile-brittle transition temperature. The change in toughness on tempering is sensitive to the carbon content of the alloy and perhaps also to the content of other interstitials. Figure 2 contains superimposed plots of the Charpy impact energy ($C_v$) and plane-strain fracture toughness ($K_{IC}$) of 9Ni steel containing ~0.06C in the quenched (Q) and quench-and-tempered (QT) conditions.[6] The intercritical temper accomplishes two changes in this case; it lowers the ductile-brittle transition temperature (DBTT) and it raises the "shelf" toughness of the alloy tested above the DBTT. Figure 3 shows a

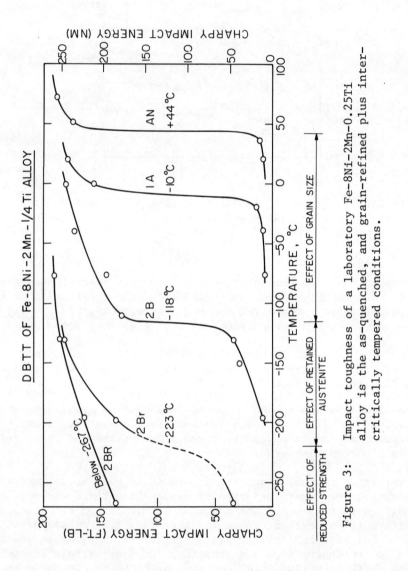

Figure 3: Impact toughness of a laboratory Fe-8Ni-2Mn-0.25Ti alloy is the as-quenched, and grain-refined plus intercritically tempered conditions.

comparable plot for a research Fe-8Ni-2Mn-0.25Ti alloy[7] which was
"gettered" of interstitials through the addition of titanium and
given a cyclic initial heat treatment to refine its grain size before tempering. The intercritical temper also lowers the DBTT in
this case, but does not sensibly affect the shelf toughness above
the DBTT. These results are typical; a suitable intercritical
temper always seems to have a beneficial effect on the ductile-to-
brittle transition, but influences toughness above the DBTT (at a
given strength level) only if the alloy contains a reasonable concentration of free interstitials.

The lowering of the ductile-brittle transition temperature on
intercritical tempering depends on the time and temperature of
tempering as well as on alloy composition.[8,9] Figure 4 illustrates
the effect of tempering temperature on the toughness of 9Ni steel
at 77°K. Optimum results are obtained when the tempering temperature is high enough to obtain a reasonable concentration of precipitated austenite, but low enough to insure that the austenite
is rich in solute. Figure 5 illustrates the effect of tempering
time at given temperature on toughness at 77°K. The toughness
increases to a maximum value, but may eventually decrease if a
very long tempering time is used.

In interpreting these and other consequences of intercritical
tempering it is usually assumed that the dominant effects are due
to the retention of the austenite precipitated during the tempering
step, though it should be kept in mind that the simultaneous tempering of martensite may also be important. The influence of retained
austenite on toughness is a subject which has not been fully resolved.
At the risk of some over-simplification, however, the results of recent research may be used to phrase a reasonably self-consistent
interpretation.

The dominant influence of retained austenite appears to have two
sources: it serves to getter deleterious interstitial species and
it acts to refine the effective grain size of the alloy. The former
effect is a dominant cause of the increased shelf toughness of carbon-
containing alloys which have been intercritically tempered; the latter effect is a dominant cause of the decrease in the DBTT.

The carbon present in a martensitic steel may be contained as a
super-saturation of free carbon in the matrix, if the steel has been
rapidly quenched, or it may be gathered into clusters or carbide
precipitates, if the steel had been more slowly cooled or tempered.
In the former case the carbon is expected to form atmospheres along
dislocations, inhibiting their mobility and hence decreasing toughness. In the latter case the carbides are preferential sites for
void nucleation, with the result that ductile crack propagation is
easier than it would otherwise be. Carbon is an austenite-stabilizing

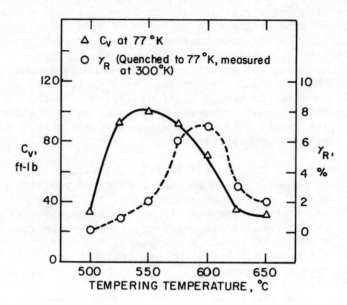

Figure 4: Impact toughness at 77°K and retained austenite content in 9Ni steel as a function of intercritical tempering temperature.

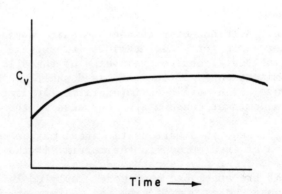

Figure 5: Schematic drawing illustrating the change in changing impact toughness at 77°K with tempering at 550°C.

# HEAT TREATMENT OF STEELS

element, and is expected to aggregate in solution in the austenite during its precipitation, hence making the matrix relatively lean in carbon and relatively tough. The process can be observed directly when the steel contains a distribution of carbide precipitates prior to tempering. The composite Figure 6 shows the progressive dissolution of carbides as austenite is introduced during the intercritical tempering of a 6Ni-0.06C alloy.

The role of retained austenite in grain-refining martensitic alloys is more subtle, and is only now beginning to be understood.[10]

The steels which are normally given an intercritical temper to improve their low temperature properties are Fe-Ni and Fe-Mn alloys, which typically form a lath-martensite structure on quenching. This structure is illustrated by the optical micrograph shown in Figure 7 and the transmission electron mirograph shown in Figure 8. The fundamental element of the microstructure is a lath of dislocated martensite. As shown in the micrographs, these laths tend to be organized into packets of parallel laths; each prior austenite grain gives rise to one or a few such packets. Crystallographic analysis of the packet shows that, while packets are not always perfect, the laths within a packet tend to have only a slight misorientation with respect to one another. Laths in a packet hence tend to share common crystallographic planes, in particular, as illustrated in Figure 9, the (100) plane which is a common cleavage plane in bcc iron. Scanning fractographic analysis of Fe-Ni steels which have failed by cleavage below the DBTT indicates that local fracture often occurs through the cooperation cleavage of the aligned laths within a packet along a common (100) plane. The effective grain size of the alloy, insofar as cleavage fracture is concerned, hence tends to be the packet size of the alloy rather than the lath size. Since the ductile-to-brittle transition in martensitic Fe-Ni and Fe-Mn alloys is usually associated with (and attributable to) a change in fracture mode from ductile rupture to quasi-cleavage, a significant decrease in the DBTT is expected if cleavage fracture is made more difficult by decreasing the packet size or by destroying the preferential alignment of laths within packets.

The decomposition of martensite packets through intercritical tempering is illustrated in Figure 10, which shows the preferential nucleation and retention of austenite along lath boundaries. Hence the martensite packets are decomposed in the as-cooled state. To insure resistance to cooperative cleavage during low-temperature testing, however, the decomposition of the martensite packet must be preserved during the test.

One of the more surprising results of recent research on the effect of retained austenite on toughness is that the austenite is almost never preserved during low temperature testing.[11,12] A virtually complete transformation of austenite is found during low

DISSOLUTION OF CARBIDE

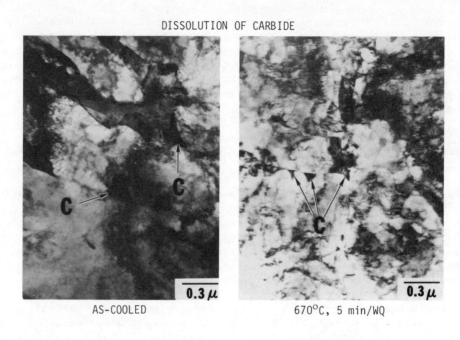

AS-COOLED  670°C, 5 min/WQ

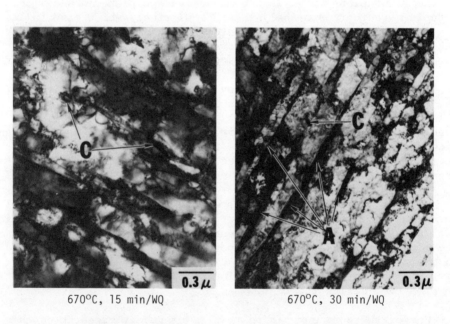

670°C, 15 min/WQ  670°C, 30 min/WQ

Figure 6: Sequence of transmission electron micrographs showing the progressive dissolution of carbides with austenite precipitation during tempering of 6Ni steel.

# HEAT TREATMENT OF STEELS

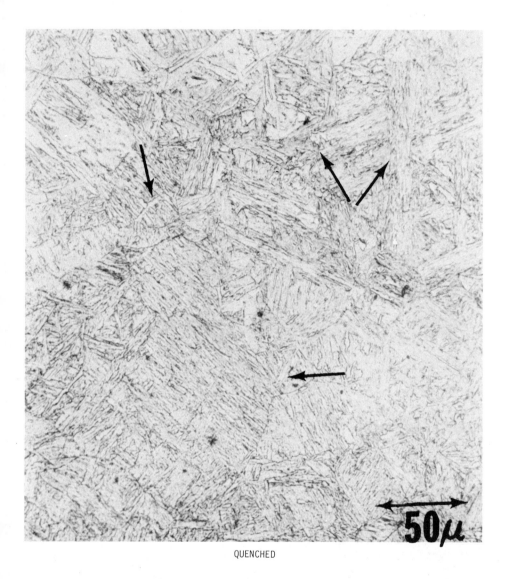

QUENCHED

Figure 7: Optical micrograph drawing the structure of martensite in 6Ni steel.

Figure 8: Transmission electron micrograph showing typical structure of dislocated martensite in quenched 9Ni steel.

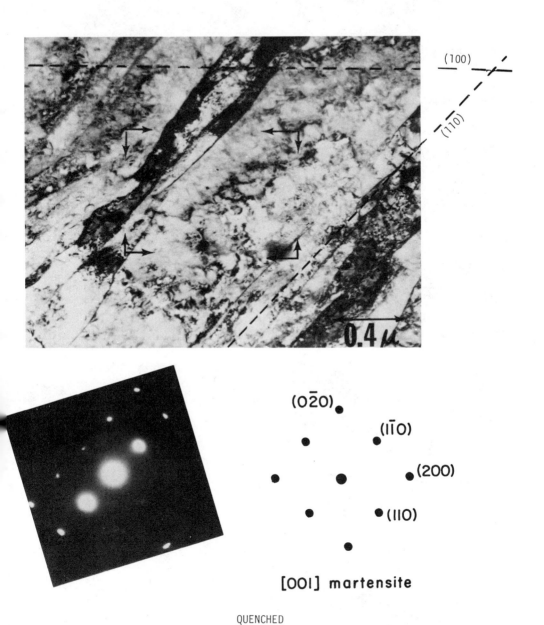

Figure 9: Transmission electron micrograph of a packet of lath martensite in 6Ni steel showing the common (100) cleavage plane.

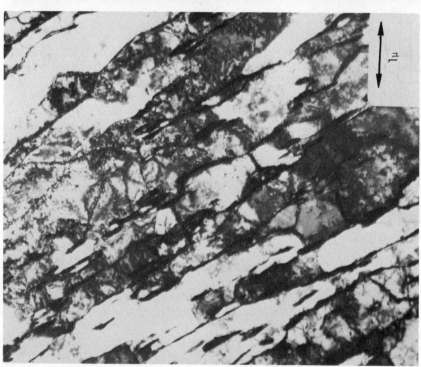

Figure 10: Transmission electron micrograph of a martensite packet in 9Ni steel showing the retention of austenite along the lath boundaries.

temperature fracture of all steels studied to date (particularly 6Ni and 9Ni cryogenic steels) even when these are tested above the DBTT. This conclusion is documented in Figures 11 and 12. Figure 11 is a set of Mössbauer spectroscopy data taken from a specimen of 9Ni steel broken at 77°K. No measurable austenite is found in the fracture surface (upper curve) although a significant concentration of austenite is present in undeformed portions of the specimen (lower curve). Figure 12 is a profile transmission electron micrograph showing a section of the ductile fracture surface of this sample. No austenite is detected in the diffraction pattern, though regions resembling retained austenite are seen in the substructure. These have apparently transformed to a highly dislocated, ductile martensite. Associated studies[10] using tensile tests suggest that the austenite is transformed at a very early stage of deformation. The influence of precipitated austenite on toughness hence involves the properties of the austenite only indirectly, through the nature of its transformation products.

The austenite precipitation during intercritical tempering is transformed prior to fracture in two stages: a portion of the austenite reverts to martensite during cooling, and the remainder is mechanically transformed during testing. Recent research has shown that these two different transformation paths have decidedly different consequences, which are illustrated in Figures 13a and 13b. Figure 13a shows a sample of 6Ni steel in which austenite nucleated along lath boundaries, as in the 9Ni sample shown in Figure 10, but reverted to martensite on cooling. The transformation has an apparently pervasive memory, so that the thermally-reverted austenite regenerates the martensite variant which gave it birth. Lath alignment within the packet is re-established, and no effective grain refinement is achieved. Figure 13b shows a sample of 6Ni steel in which the austenite reverted to martensite during mechanical deformation. In this case the memory effect is apparently over-ridden by mechanical factors, and the austenite transforms to a variant compatible with the local mechanical load. The crystallographic decomposition of the martensite packet is hence preserved after the re-transformation of austenite, with a consequent lowering of the ductile-brittle transition.

The above considerations suggest that intercritical tempering will be most beneficial when the austenite is finely distributed, so as to cause the maximum decomposition of the prior structure, and when it is stable with respect to thermally-induced transformation on cooling to the test temperature. The thermal stability of the austenite will be improved if the austenite is rich in solute and relatively small in particle size. The tempering temperature should thus be relatively low and the tempering time not too long, in keeping with practical experience. A fine distribution of austenite and a reasonable kinetics of austenite precipitation is promoted

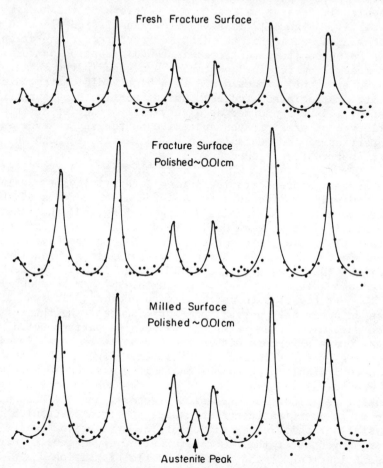

Figure 11: Mossbauer spectra taken from a Charpy specimen of 9Ni steel tested at 77°K. The upper curve is from the fracture surface; the second is from an etched surface. Neither shows the austenite peak present in the lower spectra, taken from an undeformed portion of the specimen.

# HEAT TREATMENT OF STEELS

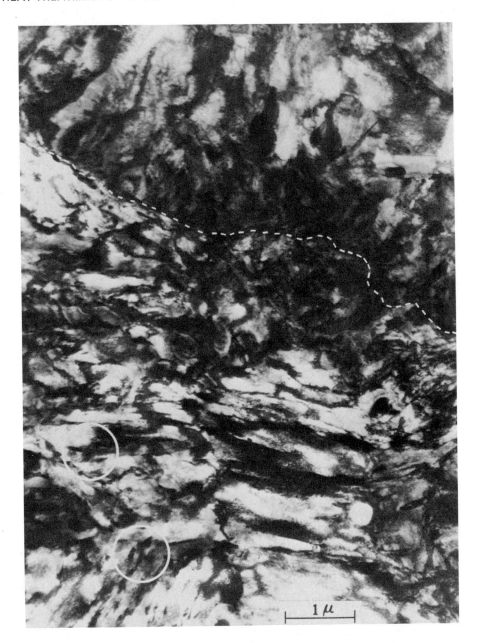

Figure 12: A profile transmission electron micrograph of a ductile fracture surface in 9Ni steel broken at $77°K$. The fracture surface is indicated. No austenite is found near the fracture surface, though regions morphologically resembling retained austenite (e.g., the circled region) are common.

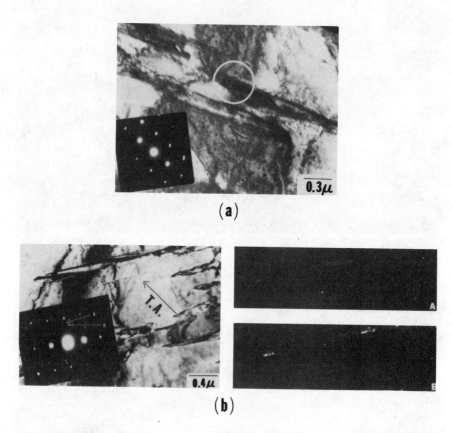

Figure 13: Comparative transmission electron micrographs showing the differing microstructural consequences of thermally and mechanically-transformed austenite in 6Ni steel. (a) After thermal transformation of austenite: the crystallographic alignment of laths within a packet is re-established. (b) After mechanical transformation at $77°K$: the crystallographic alignment of laths within the packet is broken up.

by high solute content (Ni or Mn). In 9Ni steel, the solute content is sufficiently high that a useful austenite distribution can be obtained by tempering the as-quenched alloy. Commercial 9Ni steel is hence often used in the QT condition. Less highly alloyed steels, for example, commercial 5-6Ni steels and the recently developed 5Mn composition, require prior heat treatment to achieve a suitable final structure, as will be discussed further below.

### III. INTERCRITICAL ANNEALING

If a steel is initially quenched to form dislocated martensite, and then re-heated to a point in the upper portion of the two-phase $\alpha + \gamma$ region an extensive decomposition will occur forming a high volume fraction of austenite. At the same time the residual martensite is extensively tempered, and, at least in the case of Fe-Ni steels, takes on what is essentially a polygonized ferrite structure. When the alloy is returned to room temperature most (though not necessarily all) of the precipitated austenite reverts to martensite, creating what is predominantly a two-component microstructural mixture of fresh martensite and tempered ferrite.

An example of an intercritical annealed microstructure is shown in Figure 14. The alloy shown here is a 6Ni cryogenic steel. The highly dislocated regions within the structure are fresh martensite, formed by reversion of precipitated austenite on cooling to room temperature. The intermediate regions of relatively low dislocation content are tempered martensite, which may be reasonably well described as polygonized ferrite. Some retained austenite is also present, but is not included in the micrograph. As mentioned in the previous section, there is a pronounced memory to the martensite reversion reaction, and the martensite particles tend to be very close in crystallographic orientation to the ferrite matrix. The martensite-ferrite interface appears coherent and clean, though the boundary between the highly dislocated martensite and the surrounding ferrite sometimes shows an apparent "leakage" of dislocations from the martensite, presumably to accommodate the transformation strain.

Intercritical annealing is used as one step in the thermal processing of cryogenic steels, as will be discussed below. Its principal application, however, is in the processing of the so-called "dual phase" steels[13] which are now under extensive, world-wide development for automotive applications.

The impetus for the development of dual-phase steels has its source in the increasing need for structural materials which permit weight reduction in automobiles without materially sacrificing performance or dramatically increasing manufacturing costs. The principal requirements for such an alloy call for high total tensile

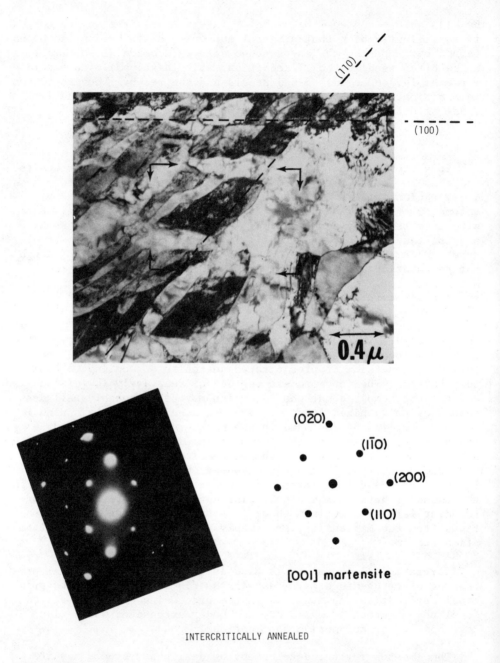

Figure 14: Structure of 6Ni steel after intercritical annealing showing highly dislocated regions representative of re-transformed austenite and regions of low dislocation density representative of tempered ferrite.

elongation, as a measure of formability, and high tensile strength, as a measure of fatigue and crush resistance.[16] It has become common to combine these criterial into a single figure of merit: the product of ultimate tensile strength and the total elongation. The steels which have the highest values of this "Figure of Merit" are the "dual phase" steels, which, as illustrated in Figure 15, offer a substantial advantage with respect to earlier automotive steels.

Despite the recent introduction of these new steels, whose publication history begins from the work of Hayami and Furukawa[13] in 1975 and of Rashid,[17] published in the next year, their evident commercial potential has led to an extensive body of research in a great many laboratories, with a consequence that varieties of dual-phase steels having a wide range of compositions and treatments have been introduced, and the metallurgy of the alloys has rapidly become complex and sophisticated. Nonetheless the fundamental step in the heat treatment of the most common alloy remains a straightforward intercritical annealing, and many of the most important properties of these alloys can be discussed in terms of the consequences of intercritical annealing for microstructure.

Figure 16 contains a schematic of the heat treatment of an Fe-Si-C dual-phase steel taken from work by Koo and Thomas.[18] The heat treatment involves a quenching step followed by an intercritical annealing. The product microstructure is shown in Figure 17. It will be seen to resemble that obtained in intercritically annealed Fe-6Ni steel presented in Figure 14. During the intercritical annealing step long islands of austenite form along the lath boundaries of the prior martensite. During subsequent cooling these revert into martensite, giving rise to the elongated regions of dislocated phase apparent in the micrograph. The highly dislocated regions are surrounded by a matrix of tempered ferrite. The interface between the ferrite and martensite is clean and coherent as illustrated by the lattice image of the interface[18] presented in Figure 18. As indicated in the figure, direct measurement of the local lattice parameter of the martensite phase shows a slight lattice expansion, which constitutes evidence for the expected accumulation of carbon in the martensite, a result of segregation of carbon to austenite during annealing. The transformation of precipitated austenite to martensite during cooling is not entirely complete; a small fraction of austenite is retained within the martensite phase. However, available evidence would suggest that this retained austenite has at best a secondary influence on the resulting properties of the steel.

As Davies[19] has emphasized, the ultimate tensile strength of a typical dual-phase steel is a nearly linear function of the volume fraction of martensite present and can be well estimated from a law of mixtures calculation. The exceptional ductility of dual-phase

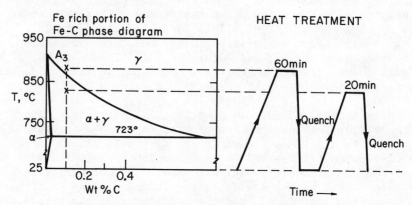

Figure 16: Schematic diagram of the heat treatment of dual phase steel (after Koo and Thomas[18]).

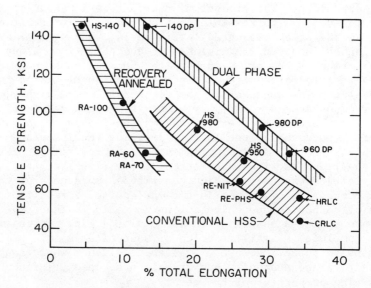

Figure 15: Tensile strength as a function of total elongation for several commercially-available automotive steels, illustrating the superiority of the dual-phase grades. (After Davies and Magee[16]).

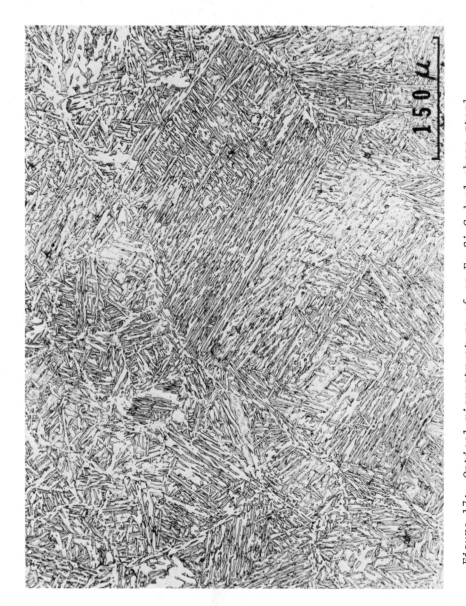

Figure 17: Optical microstructure of an $Fe_{18}Si-C$ dual phase steel after intercritical annealing.

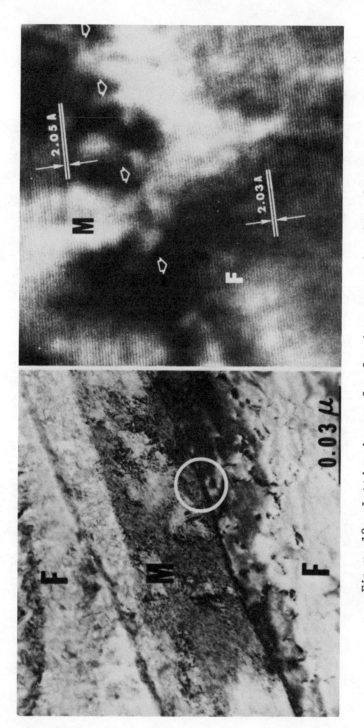

Figure 18: Lattice image of a ferrite-martensite interface in Fe-Si-C dual phase steel showing the coherence of the boundary and the lattice expansion within the martensite, presumably due to increased carbon content.

steels is more subtle in its origin and remains the subject of intensive research. The interpretation of ductility which appears to be emerging from this research may be briefly stated as follows.

The stress strain curve of a typical dual-phase steel is shown in Figure 19, and illustrates that the bulk of the total elongation occurs prior to sample necking. The point at which the sample necks in tension can be predicted from its work hardening rate $(d\sigma/d\varepsilon)$ from the Considerè criterion $d\sigma/d\varepsilon = \sigma$. Hence the elongation prior to plastic instability is equal to that which occurs before the work hardening rate falls to equal the applied stress. The sustained high work hardening rate of dual-phase alloys appears to arise from two factors: the microstructure consists of an intimate mixture of relatively soft and relatively hard components, and both components are plastically deformable. As the dual-phase steel is loaded the initial deformation is largely confined to the ferrite phase, which is relatively soft and ductile due both to its relatively low average dislocation density and to a relatively low interstitial content as a consequence of segregation of interstitials to the precipitated austenite during annealing. Since the deformation of this soft ferrite is constrained by adjacent regions of hard phase the rate of dislocation multiplication within it, and hence its work hardening rate, will be rather high. The consequence is that one achieves a significant amount of plastic elongation before the deformation of the martensite has begun to occur to any significant degree. Continued work hardening of the ferrite eventually causes its strength to exceed the yield strength of the neighboring martensite, at which point the martensite begins to undergo significant plastic deformation with accompanying work-hardening, and one achieves a second increment to the tensile elongation in which the sample is deforming more or less as would be uniform phase of high dislocation density. The sum of these two regimes of plastic flow results in a high prenecking elongation.

The post-necking elongation is also promoted by the dual-phase structure. If the hard phase within the steel were an incompatible constituent, for example a large carbide, then one would expect tensile failure to occur at a relatively early stage in deformation due to decohesion between the ferrite matrix and the hard constituent, or due to cracking of the constituent itself. In dual phase steels which have been treated to achieve good interfacial cohesion between the ferrite and martensite element, however, no such decohesion is observed and the alloy exhibits a reasonable post-necking ductility as well.

## IV. AUSTENITE REVERSION

If an initially martensitic alloy is heated to a temperature well within the austenite stability field then the structure will revert to the austenite phase. If the heating is carried out at

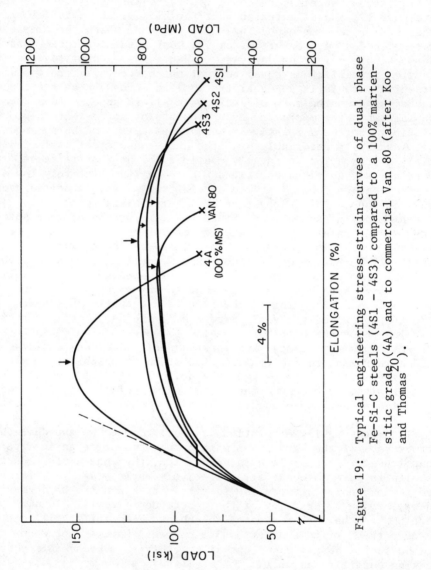

Figure 19: Typical engineering stress-strain curves of dual phase Fe-Si-C steels (4S1 - 4S3) compared to a 100% martensitic grade (4A) and to commercial Van 80 (after Koo and Thomas[20]).

a sufficiently rapid rate, or if the austenite reversion temperature is sufficiently low, then the reversion will occur through a shear mechanism which essentially reverses the martensitic transformation which created the martensite structure initially. Austenite reversion treatments have been traditionally used for many purposes in the thermal processing of steels, including normalization and initial grain refinement. The present review will emphasize two recent, and rather different applications which lead to alloys having unique or unusual properties: the transformation strengthening of austenite and the reversion treatment of maraging steels.

## A. Transformation Strengthening of Austenite

To transformation-strengthen an austenitic alloy one must begin with an alloy composition such that the martensite start temperature ($M_s$) is between room temperature and liquid nitrogen temperature ($-196°C$). The alloy may then be transformed to martensite (at least in part) by cooling in liquid nitogren. The martensitic alloy is then reverted to austenite by heating to above the austenite finish temperature ($A_f$) at a rate rapid enough to insure that the reversion reaction proceeds through a reverse shear mechanism rather than through a diffusional nucleation and growth process. When the reversion reaction is complete the alloy is cooled to room temperature. The reverted austenite is, in general, appreciably stronger than annealed austenite due principally to an increased dislocation density and to the formation of a fine substructure[21] due to the reverse shear transformation.

Initial research on the transformation strengthening of austenite was done by Krauss and Cohen,[21] who demonstrated transformation strengthening in Fe-(30-34)Ni alloys, and obtained an increase in yield strength from approximately 30 ksi to approximately 60 ksi after a $\gamma \rightarrow \alpha' \rightarrow \gamma$ reversion cycle. In further work Koppenaal[22] obtained a yield strength increase from approximately 40 ksi to approximately 105 ksi by reverting an Fe-24Ni-4Mo-0.3C alloy, and showed that the yield strength could be further increased, to ~160 ksi. In this case the strengthening is presumably due to the simultaneous influence of transformation induced defects and carbide precipitates formed during heating.

While the work cited above suggested that transformation strengthening could provide a useful alternative to thermomechanical treatment in processing austenitic steels, the approach suffered from the dual disadvantages that the strengths obtained were not exceptional in comparison to those obtainable through deformation processing or precipitation hardening from austenitic steels, and that the heat treatment employed required a very rapid heating, impractical in alloys to be used in other than very thin sections. These drawbacks appear to have been overcome in work by Jin et al.[23] which combined precipitation hardening with transformation strengthening to achieve austenite alloys having yield strengths in excess of 180 ksi, using heating rates which should be compatible with the processing of relatively thick sections.

The alloy most extensively studied by Jin et al. had composition Fe-33Ni-3Ti and was processed according to the heat treatment cycle shown in Figure 20. The evolution of microstructure during heat treatment is illustrated by the sequences of optical micrographs shown in Figure 21. In the annealed condition the alloy has a large grain size and contains a low density of internal dislocations. Ausaging at 720°C for four hours introduces a uniform dispersion of fine, spherical $\gamma(Ni_3Ti)$ precipitates, which are coherent with the matrix and contribute a substantial precipitation hardening. The yield strength of the alloy in the annealed condition is 50 ksi; after ausaging the yield strength increases to 144 ksi. The ausaging has the second consequence that the martensite start temperature of the alloy increases monotonically during aging due to depletion of nickel from the matrix. The $M_s$ temperature, which is initially below liquid nitrogen temperature, increases to approximately -70°C at the end of the ausaging treatment. An optical micrograph of the ausaged structure is shown in Figure 21a. A transmission electron micrograph of the ausaged structure showing the distribution of the ' fine precipitates is given in Figure 22a.

Because of the increase in the $M_s$ temperature during ausaging, the alloy could subsequently be transformed to martensite by refrigeration in liquid nitrogen. Cooling in liquid nitrogen produces a microstructure which contains ~60% martensite phase. An optical micrograph showing the resulting dense distribution of martensite plates is presented in Figure 21b. A corresponding transmission electron micrograph which reveals that the martensite is internally twinned, is shown in Figure 22b.

Following refrigeration in liquid nitrogen, the alloy could be reverted to austenite by heating to 720°C at a controlled heating rate of ~8°C/minute. The reheating accomplishes a complete reversion of the martensite phase. After subsequent cooling to room temperature, the reverted austenite has a yield strength of 173 ksi, indicating a significant increment in strength due to the transformation. An optical micrograph of the reverted alloy is shown in Figure 21C. While this alloy is entirely austenitic, there remain martensite-like features within the microstructure, which are ghosts of the martensite formed on refrigeration in liquid nitrogen. Transmission electron microscopic studies, as illustrated in Figure 22C, show that these martensite-like features are regions of very high dislocation density, presumably created by the reverse shear transformation of the martensite particles. The structure of the alloy hence consists of a mixture of highly dislocated austenite, which

# HEAT TREATMENT OF STEELS

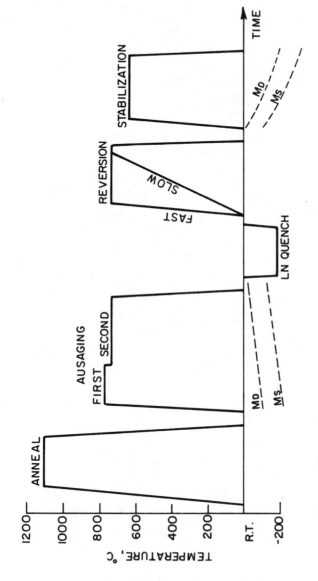

Figure 20: Schematic diagram of the processing sequence used to strengthen Fe-Ni alloys.

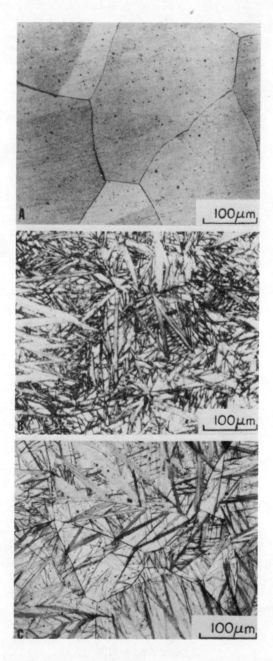

Figure 21: Optical micrographs showing the evolution of the microstructure during transformation strengthening of Fe-33Ni-3Ti: (a) as-ausaged; (b) after cooling in liquid nitrogen; (c) after reversion to austenite.

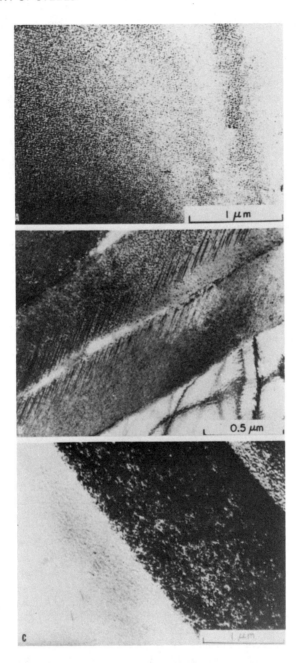

Figure 22: Transmission electron micrographs showing the evolution of the microstructure during transformation strengthening of Fe-33Ni-3Ti: (a) as-ausaged; (b) after cooling in liquid nitrogen; (c) after reversion to austenite.

arose from the retransformation of the martensite phase, and very soft dislocation-free austenite which corresponds to the untransformed matrix.

Very high strength austenitic alloys can hence be obtained by transformation strengthening a matrix which has previously been precipitation hardened. The use of a precipitation-hardened alloy has the additional advantage that the transformation-induced defects are very stable to annealing at elevated temperature. Figure 23 shows a transmission electron micrograph of a region of the austenite which was reverted and held at 720°C for 7 hours. There is only a very slight deterioration in yield strength after this extended holding. Transmission electron micrographic studies show that the highly dislocated region remained intact and stable during annealing. There is, however, some evident fuzziness in the periphery of these regions after 7 hours of annealing, indicating a gradual leakage of transformation-induced dislocations into the surrounding austenite phase.

The disadvantage of transformation hardening as a means of strengthening austenite is that the treatment is only applicable to alloys whose compositions had been so adjusted that the $M_s$ temperatures lie in an appropriate range below room temperature. It is also important that after transformation strengthening has been carried out the product austenite has sufficient mechanical stability that it does not catastrophically transform to martensite under elastic load, or the strength of the alloy will be low and determined by the stress which induces martensite transformation rather than the stress which causes plastic yielding.

### B. Maraging Steels

The reversion treatment has also been used to achieve unusually high strength-toughness combinations in conventional maraging steels. The work described here was carried out by Jin and Morris, and is described in reference (24).

The so-called maraging steels are alloys of iron, nickel, cobalt, molybdenum and titanium which are low in carbon and known to have very good strength-toughness characteristics. The alloys are usually processed by quenching to form martensite followed by normal tempering to bring out precipitation hardening species, which may be Mo clusters, Ni-Mo precipitates, or other intermetallic compounds. A further improvement in the strength-toughness characteristics of these alloys would, however, be desirable, and has been sought by investigators for some time.

Early research toward improved strength-toughness combinations in maraging steels concentrated on attempts to improve the toughness of the steels at a given strength level by introducing an

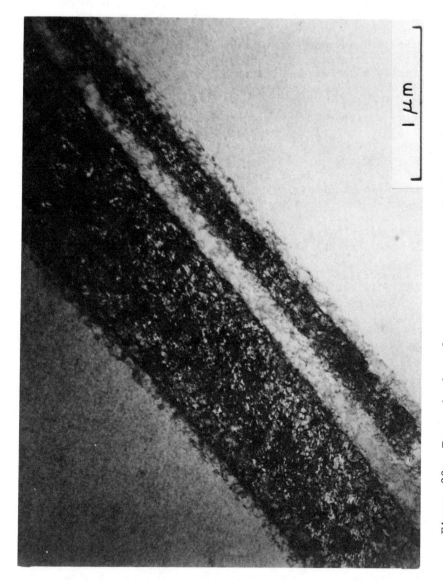

Figure 23: Transmission electron micrograph showing retention of transformation-induced defects in Fe-33Ni-3Ti after seven hours holding at 720°C.

austenite phase through an intercritical tempering treatment. While researchers have reported toughness improvements in maraging steels associated with the introduction of austenite, a closer examination of this data[25] reveals that either the apparent toughness improvement results from a loss of alloy strength or depends on toughening mechanisms other than the introduction of austenite. A specific investigation of the change in the strength-toughness characteristic of the 250 grade maraging steel on the introduction of austenite by intercritical tempering[25] shows that the austenite has essentially no effect on the strength-toughness characteristic. This result is, of course, consistent with results obtained from research on cryogenic steels, and reported in section II, when these steels are tested above their ductile-brittle transition temperatures.

A more successful approach to improve the toughness of maraging steels employs a reversion treatment, and has been used by Jin and Morris[24] and also, implicitly, by Antolovich et al.[26] In this approach the steels are first maraged to establish a precipitate distribution, rapidly reverted to austenite and retransformed to martensite to establish a very high dislocation density, and finally re-aged to stabilize the high dislocation density and early yielding.

The microstructure of a 250 grade maraging steel aged at 550°C for five hours is shown in Figure 24. The maraged precipitates in the martensite matrix are clearly seen. These proved to be primarily $Ni_3Mo$ with some admixture of $Ni_3Ti$. When the martensite with embedded precipitates is subjected to a rapid reversion to austenite and retransformed to martensite an extremely high dislocation density is established. This dislocation density is illustrated in Figure 25. Also present in the microstructure are discrete islands which appear as large white particles in the micrograph. These are retained austenite particles, and make up ~10-15% of the alloy.

From the appearance of the microstructure shown in Figure 25, particularly with respect to its very high dislocation density, one would expect that the alloy would have a very high yield strength. In fact, the yield strength of the maraging steel decreases slightly on reversion treatment. An analysis of the stress-strain curve for the alloy suggests that the decreased yield strength is due to an incipient yielding at a rather low stress level. This incipient yielding may be attributed to the very high density of fresh dislocations, some of which are apparently mobile under rather low stresses. To pin these dislocations and increase the strength, the alloy was given a second low temperature aging at 380°C. This secondary low-temperature aging adds an increment of 30 to 50 ksi in the yield strength of the alloy.

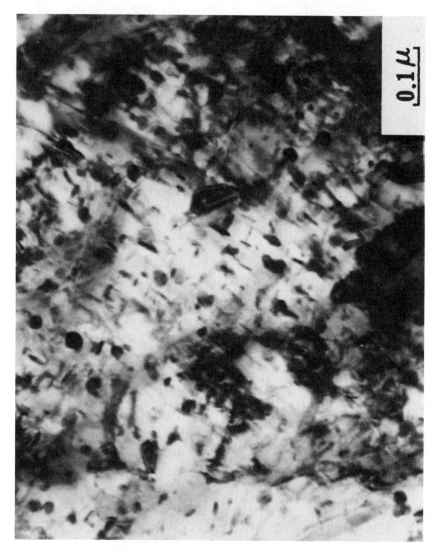

Figure 24: Transmission electron micrograph of 250 grade maraging steel aged at 550°C for five hours.

Figure 25: Transmission electron micrograph of 250 grade maraging steel aged at 550°C for five hours, then given a reversion treatment.

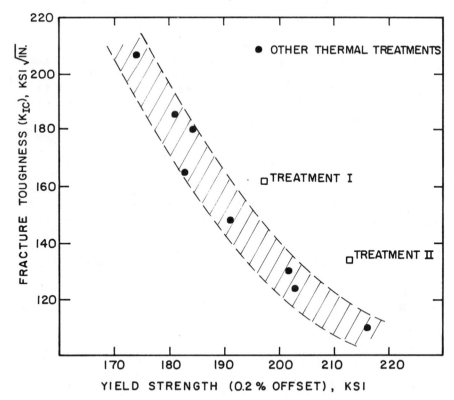

Figure 26: Comparison of strength-toughness characteristic of 250 grade maraging steels given reversion treatment (I,II) with that of the same steel after normal thermal treatments.

The strength-toughness data for 250 grade maraging steels processed by maraging plus reverse transformation plus secondary aging are compared to the strength-toughness characteristics of alloys processed through normal maraging in Figure 26. As can be seen from the figure, the reverse transformation does produce an improvement in the strength-toughness characteristic. A fracture toughness in the neighborhood of 160 ksi√in at a yield strength of 200 ksi was achieved through the reversion treatment. This combination of strength and toughness exceeds any previously published for the 250 grade maraging steel.

It would appear from these results that reversion treatments can be successfully used to improve the strength-toughness characteristics of high strength structural steels intended for use near room temperature.

## V. COMBINATIONS OF THERMAL TREATMENTS

The characteristic heat treatments which have been discussed individually in preceding sections can also be used in combination to establish alloys having particularly desirable properties. Cycling heat treatments, for example, are commonly used in the processing of specialty alloys intended for use under extreme conditions. Examples include the so-called QLT, 2B, and 2BT treatments which have been found useful in the processing of steels intended for cryogenic use.

Low nickel cryogenic steels, including both the 5Ni "Cryonic 5" produced by the Armco Steel Co., and the 5.5Ni cryogenic steel produced by the Nippon Steel Co., are processed through a treatment which, in the case of the Nippon alloy, is known as the QLT treatment.[27] The processing sequence consists of a quenching followed by an intercritical annealing followed by an intercritical tempering to introduce retained austenite. It is found that while the intercritical tempering does not itself improve cryogenic properties, it substantially improves the cryogenic toughnesses which can be obtained by a subsequent intercritical tempering. The advantage introduced by the intermediate intercritical anneal is not yet fully understood. Our preliminary research indicates, however, that alloys which have been intercritically annealed will establish a much finer and more uniform distribution of precipitated austenite on subsequent intercritical tempering. The reason is believed to lie in the nickel segregation associated with the intercritical annealing step, which provides a density of sites relatively high in nickel content at which austenite can easily precipitate during subsequent intercritical tempering.

The 2B cycling treatment is diagrammed in Figure 27. It consists of a four step thermal cycling treatment in which austenite reversions are alternated with intercritical annealing treatments. Its purpose is to grain refine the alloy matrix, and it has been found successful in reducing the grain size of Fe-Ni[28] and Fe-Mn[29,5] steels to ~1 micron. The 2B treatment was initially used to grain refine Fe-12Ni-0.25Ti alloys so that they would retain high toughness with good structural strength in liquid helium. It was found that 12Ni steel processed through the 2B treatment would retain toughnesses greater than 200 ksi in at yield strength near 200 ksi at 4.2 - 6°K.[5]

The 2B treatment by itself, however, has not been found sufficient to toughen alloys containing carbon for low temperature use. In carbon-containing alloys it is necessary to add an intercritical tempering after the 2B treatment has been used to establish a very fine grain structure. The resulting 2B treatment was shown to be sufficient[6] to establish an excellent combination of strength and

# HEAT TREATMENT OF STEELS

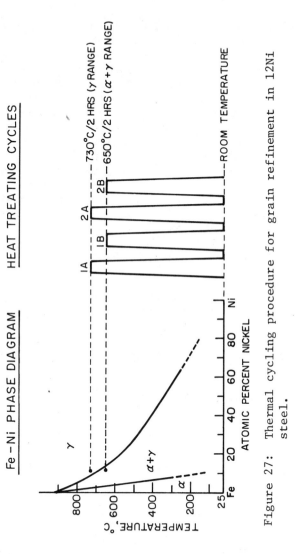

Figure 27: Thermal cycling procedure for grain refinement in 12Ni steel.

toughness in commercial 9Ni steel at 4-6°K. In very recent research this treatment has also been shown to establish a good combination of strength and toughness in Ni-free Fe-5Mn alloys[5] at liquid nitrogen temperature.

## VI. CONCLUSION

The examples described in this paper were put forward for two reasons: To illustrate the significant advances in properties of steel which have been obtained recently through the sophisticated use of heat treatment and to simultaneously illustrate the fundamental simplicity of the heat treatments employed. Despite the simplicity of these heat treatments, and the emerging general picture of the associated changes in the microstructures of steels, it is nonetheless true that many features of the response of typical alloys through these characteristic heat treatments remain poorly understood. It may be expected that still further advances and improvements in the properties of steels will be obtained as continued research establishes a better fundamental understanding of the response of steels to heat treatment which would involve some reversion to the austenite phase and permit a more precise control of the microstructures obtained.

### Acknowledgement

The preparation of this paper was supported by the Office of Naval Research under Contract N00014-75-C-0154. Research reported here draws on work supported by the Office of Naval Research, by the Division of Basic Energy Sciences of the U.S. Department of Energy under Contract W7405-ENG-48; by the Electric Power Research Institute under Contract RP636-2 and by the Air Force Materials Laboratory under Contract F33615-73-C-5100. The authors appreciated the assistance of J. Y. Koo and G. Thomas in providing illustrative micrographs of dual-phase steels.

### References

1. C. W. Marshall, R. F. Heheman, and A. R. Troiano, Transactions Am. Soc. Metals, 55, 135 (1962).

2. T. Ooka and K. Sugino, *J. Japan Inst. Metals*, 30, 435 (1977) (in Japanese).

3. S. Nagashima, T. Ooka, S. Sakino, H. Mimura, T. Fuzishima, S. Yano, and H. Sakurai, *Trans. I.S.I.J.*, 11, 402 (1971).

4. D. Sarno, J. Bruner, Armco Steel Corp., Private Communication.

5. M. Niikura and J. W. Morris, Jr., *Met. Trans.* (submitted).

6. C. K. Syn, S. Jin and J. W. Morris, Jr., Met. Trans., 7A, 1827 (1976).

7. S. Jin, S. K. Hwang, and J. W. Morris, Jr., Met. Trans., 6A, 1721 (1975).

8. T. Ooka, H. Mimura, S. Yano, K. Sugino, and T. Toisumi, J. Japan Inst. Metals, 30, 442 (1966) (in Japanese).

9. S. K. Hwang, S. Jin, and J. W. Morris, Jr., Met. Trans., 6A, 2015 (1975).

10. J. W. Morris, Jr., C. K. Syn, J. I. Kim, and B. Fultz., Proceedings - ICOMAT-79, Cambridge, Mass., June, 1979.

11. C. K. Syn, B. Fultz, and J. W. Morris, Jr., Met. Trans., 9A, 1635 (1978).

12. B. Fultz, M.S. thesis, Dept. Materials Science and Mineral Engineering, U. California, Berkeley, 1978. (Lawrence Berkeley Lab., Rept. LBL-7671).

13. S. Hayami and T. Furukawa, Proceedings, Micmalloy-75, Washington, 1975, pp. 78-87.

14. Modern Developments in HSLA Formable Steels (Proceedings: AIME Symposium Chicago, 1977) A. T. Davenport, Ed. (in press).

15. Structure and Properties of Dual Phase Steels (Proceedings: AIME Symposium, New Orleans, 1979) R. C. Kot and J. W. Morris, Jr., eds., (in press).

16. R. G. Davies and C. L. Magee, in Ref. (15).

17. M. S. Rashid, SAE Preprint 760206, Feb. 1976.

18. J. Y. Koo and G. Thomas, Met. Trans. 8A, 525 (1977).

19. R. G. Davies, Met. Trans., 9A, 671 (1978).

20. J. Y. Koo and G. Thomas, in reference 14.

21. G. Krauss, Jr., & M. Cohen, Trans. TMS-AIME, 229, 1212 (1962).

22. T. J. Koppenaal, Met. Trans., 2, 1549 (1972).

23. S. Jin, J. W. Morris, Jr., Y. L. Chen, G. Thomas and R. I. Jaffe, Met. Trans., 9A, 1625 (1978).

24. S. Jin and J. W. Morris, Jr., Rept. AFML-TR-75-119, May, 1975.

25. S. Jin, S. Hwang and J. W. Morris, Jr., Met. Trans., 9A, 637 (1976).

26. S. D. Antolovich, A. Saxena, and G. R. Chanini, Met. Trans., 5, 623 (1974).

27. S. Yano, H. Sakurai, H. Mimura, N. Wakita, T. Ozawa, and K. Aoki, Trans. I.S.I.J., 13, J33, (1978).

28. S. Jin, J. W. Morris, Jr., and V. F. Zackay, Met. Trans., 6A, 191 (1975).

29. S. K. Hwang and J. W. Morris, Jr., Met. Trans. (in press).

30. S. Jin, S. K. Hwang, and J. W. Morris, Jr., Met. Trans., 6A, 1569 (1975).

INNOVATIONS IN GRINDING MATERIALS


R. A. Rowse and J. E. Patchett

Norton Research Corporation (Canada) Limited


INTRODUCTION

Grinding is an old process to accomplish stock removal of metals. Silicon carbide and alumina are the common materials used in grinding wheels and in sand paper. These materials were first made synthetically about the turn of the century. The review by Ueltz (1) describes these developments in detail.

Until recently, alumina and silicon carbide were the only synthetic abrasives made in large quantities and consequently these were used for all types of grinding applications. Today there are many types and specific abrasives should be selected for specific applications, particularly if performance and cost values are important. The development of ceramic alloys was one of the reasons which has led to specific abrasives. This development was a major innovation considering only a few years ago the industry was considered to be "mature" with no new changes likely.

In the late 1950's the abrasive industry discovered the advantages in the co-fusion of alumina and zirconia (2). It was, however, another 10 years later in 1968 before a product of considerable significance was introduced. In 1972, a second alumina-zirconia composition was developed for Coated Abrasive belts and discs and grinding wheel applications requiring low grinding forces (3). The importance of these $Al_2O_3$-$ZrO_2$ abrasive materials at their introduction and at the present in producing changes in the construction of grinding wheels and machines and operating conditions for steel conditioning is shown in Table I.

The improvements in grinding efficiencies for these combined changes in abrasives and machines is shown in Table II.

TABLE I

Machines and Grinding Wheels for Steel Conditioning

| Year | Abrasive Type | Wheel Size In Inches Diameter | Width | Wheel Speed feet/min | Grinding Force in pounds | Horsepower |
|---|---|---|---|---|---|---|
| 1950 | Fused Alumina | 16 | 2 | 9,600 | 400 | 25 |
| 1956 | Fused & Cast Alumina | 20 | 2½ | 12,000 | 700 | 40 |
| 1962 | Fused $ZrO_2$ & $Al_2O_3$ (slow cooled) | 24 | 3 | 12,000 | 900 | 75 |
| 1968 | Fused $ZrO_2$-$Al_2O_3$ (fast cooled) | 24 | 3 | 16,000 | 2,000 | 150 |
| 1977 | Fused $ZrO_2$-$Al_2O_3$ (fast cooled) | 36 | 4 | 16,000 | 6,000 | most 150-250 few 250-400 |

# INNOVATIONS IN GRINDING MATERIALS

## TABLE II

### Relative Performance Trends in Abrasives

| Year | Abrasive Type | Grinding Ratio |
|------|---------------|----------------|
| 1950 | Fused Alumina | 100 |
| 1956 | Fused and Cast Alumina | 186 |
| 1962 | Fused $Al_2O_3$-$ZrO_2$ (slow cooled) | 256 |
| 1977 | Fused $Al_2O_3$-$ZrO_2$ (fast cooled) | 586 |

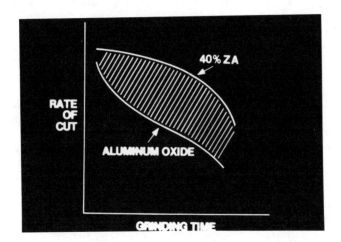

Figure 1. Typical cutting performance improvement of $Al_2O_3$-$ZrO_2$ abrasives compared to the standard $Al_2O_3$ abrasive.

The $Al_2O_3$-$ZrO_2$ composition developed for Coated Abrasive applications exhibited equally dramatic improvements over fused aluminum oxide with a quality improvement of several hundred % in nearly all applications, see Figure 1.

These advances in materials have been made possible by the introduction of new thinking in the design of research experiments and in manufacturing procedures.

What has emerged is the realization that the mixture of alumina and zirconia are true alloys and consequently can be manufactured in various ways so as to improve the properties. Furthermore, the metal alloy technology can be adopted for the oxide alloys.

## II. ALLOY ABRASIVES

Alloys of alumina and zirconia are the only material of commercial importance at present. The first of these alloys have a composition of $Al_2O_3$-25% $ZrO_2$ and when rapidly quenched and crushed to yield a strong particle becomes an extremely durable abrasive. Its applications are the surface conditioning of steel ingots. The requirements are for large metal removal rates with minimal wheel wear rates (high G ratio). These grinding wheels are subjected to very high forces and are driven by large motors — 500 horsepower. This application is growing because other means of conditioning steel ingots such as scarfing, chipping, acid cleaning are expensive and cause pollution problems. Another application is in foundries to clean castings and cut off "risers" and the like.

The second alloy composition is $Al_2O_3$-40% $ZrO_2$ and when rapidly quenched this has the cellular two phase eutectic microstructure. It has the unique and very desirable property of being able to microfracture in use at moderate to low grinding forces. This abrasive finds utility in coated abrasive belts and discs and in grinding wheels used for cut-off, heavy duty finishing and portable grinding.

With the success of the alumina-zirconia alloys many other systems have been studied by the industry and these are appearing in the patent literature such as Spinel-$ZrO_2$, ternary oxide systems, solid solution with $Al_2O_3$ and $ZrO_2$ and the like. In addition, there would seem to be a renewed interest in the various academic institutions as evidenced by the number of published papers on ceramic alloy systems (5-6). Still another active area for ceramic alloy systems has been the investigation by United Technologies Research Laboratory to utilize $Al_2O_3$-$ZrO_2$ ($Y_2O_3$) alloys and other alloys in turbine engine parts.

## III. ALLOY ABRASIVE MANUFACTURE

The abrasive industry had to develop molds for rapidly quenching the alloy abrasives. The requirements were specialized in that the melting temperatures are in the excess of 1850°C, the surface condition of the castings were not critical, heat flow from the alloy into the mold was slow and the casting must be crushed up into grit size particles.

The various methods having commercial significance have been patented and no doubt other processes will continue to be patented by the principal North American manufacturers — Norton, Carborundum, Exolon and General Abrasives and by Japanese and European companies.

# INNOVATIONS IN GRINDING MATERIALS

The technical accomplishments are large in that the quenching rate is about a thousand degrees centigrade per second for the modern versions of these alloy abrasives. Furthermore, this is not on a laboratory scale as Norton Company, as just one producer, is casting over 10,000 tons per year of alumina-zirconia alloy abrasives.

It was soon realized that the quality of the abrasive was directly related to the quality of the casting. The relationship was indeed very similar to metal alloy castings. This relationship between properties and processing and end use forced the industry to characterize very thoroughly the alloy abrasives.

## IV. CHARACTERIZATION FEATURES

The studies relating properties of the material and process changes have revealed features which have been of considerable concern to those people working with rapidly solidified oxides. Some of these features using $Al_2O_3$-$ZrO_2$ as an example are the following:

The data developed by Norton Company tend to support the following Phase Diagram: (Figure 2)

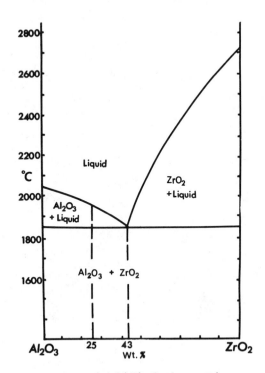

Figure 2. Simplified Phase Diagram

The composition of the two commercially available alloys are shown on this diagram. Our studies have not attempted to investigate solid solution limits. There has been no attempt to show the various polymorphs of $ZrO_2$ as this was not a primary concern. This diagram, however, is idealized in that it describes equilibrium conditions while most commercial processes should be considered non-equilibrium.

## Non-Equilibrium Conditions

At the solidification rates needed to produce very fine microstructures, non-equilibrium conditions exist. This is evident in what would appear as undercooling and by the shift in the eutectic composition to lower $ZrO_2$ levels.

Another feature of non-equilibrium solidification is the amount of metastable tetragonal $ZrO_2$ that remains in the solidified alloy. This condition has been observed by several researchers and would seem to be related in part to the crystallite size of the $ZrO_2$ (7). In some cases the crystallite size is limited to a few hundred angstroms by the eutectic rod and lamellar size. However, high tetragonal content can be obtained with lamella coarser than 300 angstroms so size is not the only explanation. Some other hypothesis have been suggested by Kuriakose and Beaudin (8). Norton Company studies have shown a strong positive correlation between the amount of tetragonal $ZrO_2$ and grinding quality. The amount of tetragonal $ZrO_2$ probably represents the combined effects of several processing steps.

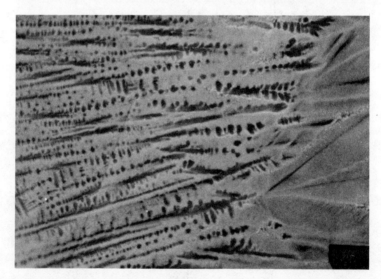

Figure 3. Dendritic "cubic" $Al_2O_3$. Chill is at the left side of the picture. A two phase cellular mixture is on the right. Bar is equal to 20 microns.

# INNOVATIONS IN GRINDING MATERIALS

Still another feature of non-equilibrium conditions is the existence of a cubic polymorph of alumina. The evidence is shown by x-ray diffraction as well as the morphology of the dendrites. (Figure 3) Such observations have been described by those doing plasma spraying of oxides (9) and is perhaps the explanation why the published $Al_2O_3$-$ZrO_2$ phase diagram by Cevales (11) shows a cubic polymorph of alumina. This cubic alumina phase produced from melts is not well known.

## Differential Growth Rates

Heat flow through the alumina zirconia alloy is slow. Consequently, those portions of the alloy in contact with the mold quench at a faster rate than the center portion of the casting. This results in a considerable range in rates of solidification and growth. In our experience, the eutectic type microstructure grows more rapidly than dendritic alpha alumina. There are also suggestions that the cubic alumina grows only at a still high rate than the two phase mixture. Thus a cross-section of the casting will show zones of very different phase assemblages. (Figure 4).

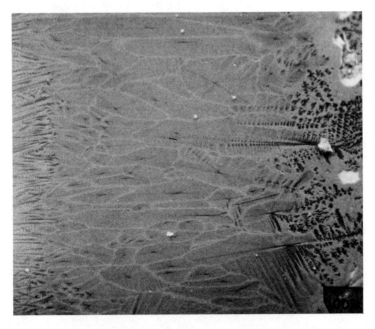

Figure 4. Cross-section with chill on the left. Fastest growth rate is "cubic" $Al_2O_3$ followed by 2 phase cellular mixture, followed by rhombic $Al_2O_3$ and 2 phase cellular mixture. Bar is equal to 100 microns.

### Microhardness

This observation on microhardness of alloy abrasives would show many similarities to that experienced in metal alloys. That is, the microhardness of alloy abrasives is clearly depended on the microstructure of the alloy. The data show the microhardness of the cells to be nearly equal to that of alpha alumina when the two phase mixture is fine.

### Quantitative Microstructure

The solidification of alloy abrasives can be expressed in the same terms used to define metal alloy systems. For example, it is possible to use the theoretical expressions for rate of solidification and with reasonable assumptions find close agreement between calculated and observed values. Similarily the lambda value — the spacing between rods or lamellae, has been found to be equally significant. The study has been most helpful in understanding heat flow from the casting into the mold and in distinguishing various castings.

In the dendritic alloys, the 25% $ZrO_2$ alloy, the secondary dendrite arm spacing has been an important measurement as this can be related to grinding quality. The appearance of S.D.A. and the change in arm spacing as affected by thickness of the casting is shown in Figures 5 and 6.

TABLE III

|  | Knoop Hardness Kg per mm$^2$ |
|---|---|
| $Al_2O_3$ | 2000 |
| $ZrO_2$ | 1200 |
| $Al_2O_3$-$ZrO_2$ (fast cooled) | 1810 |
| Cell Boundary | 1300 |

# INNOVATIONS IN GRINDING MATERIALS

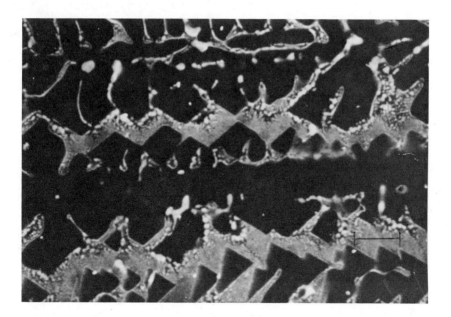

Figure 5.  Dendritic rhombic $Al_2O_3$ surrounded by two phase cellular mixture of $Al_2O_3$ and $ZrO_2$. Secondary Dendrite Arms are faceted. Bar is equal to 20 microns.

## Microfracture

The property of being able to microfracture is most important for abrasives. An idealized abrasive should wear away slowly and maintain its sharp cutting ability. A dulling action or complete fracture of the abrasive is undesirable.

A cellular alloy abrasive approaches this idealized abrasive in that the cells are much finer than the grit particles, the cell boundaries are weak and the fracture planes and the cellular structure can be directionally produced to optimize the fracturing pattern.

As in metal alloy systems the chemical impurities in the alloy abrasive tend to be concentrated in the cell boundaries. The kinds and amounts of impurities have a major influence on the microfracturing and hence grinding properties of these abrasives. These microstructural feasures relating to micro-fracture are shown in Figures 7, 8 and 9.

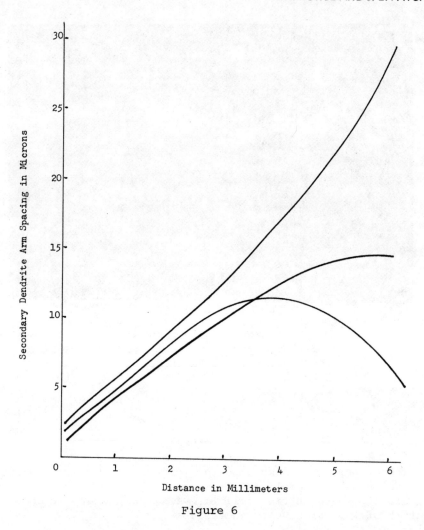

Figure 6

## V. GRINDING TESTS

It soon became apparent that improvements were needed in evaluating abrasives in order to differentiate between various materials and to relate properties to grinding performance. There are many variables in a grinding operation and it is difficult to control them. Most grinding tests produce relative data and this makes it difficult to define the results in quantitative terms. The methods of grinding and interpretation of data has been most successful when procedures developed by Coes, (11), Lindsay, (12) and Story (13) have been followed. The various tests take into consideration metal removal rate, abrasive wear rates, power usage, grinding forces and

# INNOVATIONS IN GRINDING MATERIALS

grinding fluids. In addition, several grinding conditions should be considered and for some steels their grindability has to be taken into consideration.

There is a need for better end use tests for abrasives and the need to develop laboratory tests which can more accurately predict performance for production scale operations.

## VI. SUMMARY

The discovery of alloy abrasives has drastically changed our concepts of what constitutes an abrasive material and has greatly improved the grinding operation. There are strong similarities to metal alloy systems and many aspects of the metal alloy technology can be adopted for oxide alloys. Alumina-zirconia alloy abrasives would seem to be the first commercial application but one might expect other used to be developed for oxide alloys.

## VII. ACKNOWLEDGEMENTS

The authors wish to thank Norton Company for permission to publish this article and our colleagues who have carried out research, development and production of alumina-zirconia abrasives.

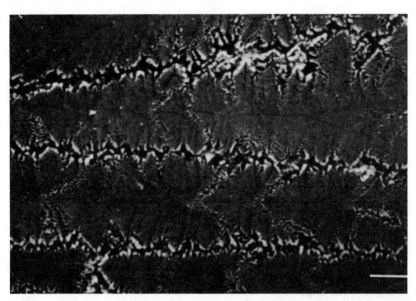

Figure 7. Directionally solidified cells of $Al_2O_3$ and $ZrO_2$. The cell boundaries are the locations for impurities and the coarsening of the $ZrO_2$. The cell boundaries are weak areas. Bar is equal to 40 microns.

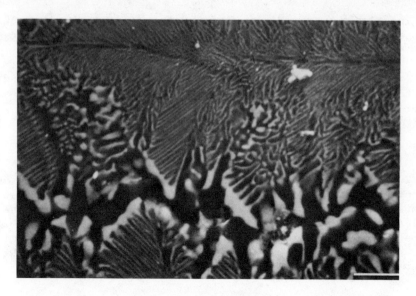

Figure 8. Enlarged area of $Al_2O_3$-$ZrO_2$ cell and cell boundary. The $ZrO_2$ lamellae at the center of the cells become much coarser in the boundaries. Impurities are also present in boundary. Bar is equal to 10 microns.

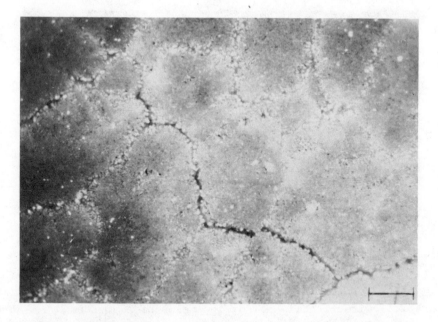

Figure 9. Cracks through the abrasive tend to follow the weak grain boundaries. This promotes a controlled microfracture. The bar is equal to 40 microns.

## REFERENCES

1. H. F. Ueltz "Abrasive Grains - Past, Present and Future", in New Developments in Grinding, ed. by M. C. Shaw, Carnegie Press, Pittsburgh, Penn. (1972).

2. D. W. Marshall and S. J. Roschuk, assigned to Norton Company, "Fused Alumina-Zirconia Abrasives", U. S. Patent 3,181,939, May 4, 1965.

3. R. A. Rowse and G. R. Watson, assigned to Norton Company, "Zirconia-Alumina Abrasive Grain and Grinding Tools", U. S. Patent 3,891,408, June, 1975.

4. L. Coes Jr., assigned to Norton Company, "Zircona-Spinel Abrasive", U. S. Patent, 3,498,769, March 1970.

5. R. L. Ashbrook, "Directionally Solidified Ceramic Eutectics", J. Am. Ceram. Soc., 60, 428-435, 1977.

6. R. E. Shepler, "Influence of Microstructure on Friability of Alumina-Zirconia Abrasives", Ph. D. Thesis, University of Florida, 1975. University Microfilms International, Ann Arbor, Michigan, 1977.

7. C. O. Hulse, et al, assigned to United Aircraft Corporation, "Directionally Solidified Refractory Oxide Eutectic", U. S. Patent, 3,761,295, September, 1973.

8. R. C. Garvie, "The Occurence of Metastable Tetragonal Zirconia as a Crystallite Size Effect", J. Phys. Chem. 69 (4), 1238-1243, 1965.

9. A. K. Kuriakose and L. J. Beaudin, "Tetragonal Zirconia in Chilled-Cast Alumina-Zirconia", Jr. Canadian Ceramic Society, 46-50, 1977.

10. A. Krauth and H. Meyer, "Modifications Produced by Quenching and the Crystal Growth in Systems Containing Zirconia", Ber. Deut. Keram. Ges., 42, 61-72, 1965.

11. G. Cevales, "Phase Equilibrium Diagram of Alumina-Zirconia and Examination of a New Phase ($Al_2O_3$)", Ber. Deut. Keram, Ges. (45), 216-219, 1968.

12. L. Coes, Jr., "Abrasives", Springer-Verlay, New York, 1971.

13. R. P. Lindsay, Variables Affecting Metal Removal and Specific Horsepower in Precision Grinding", SME paper No. MR71-269. Society of Manufacturing Engineers, Dearborn, Michigan, 1971.

14. R. W. Story, Unpublished Information, Coated Abrasive Division, Norton Company, Troy, New York.

RECENT ADVANCES IN GRINDING


P. Guenther Werner

Massachusetts Institute of Technology
Cambridge, Massachusetts


INTRODUCTION

Two decades ago grinding was almost totally applied as a finishing process working at very low removal rates and correspondingly low productivity. At those times grinding was used mainly because other methods could not provide the required geometrical accuracy or could not cut hardened materials at all. In the recent years, however, the situation has changed substantially. Still used as the dominant finishing process, grinding has been developed such that increased removal rates at constant or increased accuracy became possible. Also high-efficiency grinding of difficult-to-machine materials like stainless steels, tool and high speed steels, titanium and nickel-cobalt-base alloys, is performed today at a larger scale.

This development has been anticipated by a representative group of manufacturing experts assessing the undeveloped potention of the grinding process a few years ago (1). Figure 1 reflects the group opinion for some selected statements. The majority, for example, expected new abrasive materials and exact analytical methods to describe the grinding process to become effective between 1980 and 1990. The substitution of other manufacturing processes by grinding is expected to increase steadily; grinding itself, however, is expected to be never substituted by other manufacturing processes completely. Other statements were evaluated in a more conservative way: About 30% of the questioned experts think that grinding wheel speeds applied in production will never be increased to 300 m/s (60000 sfpm), and 25% believe that grinding will never cover 50% of all finishing operations.

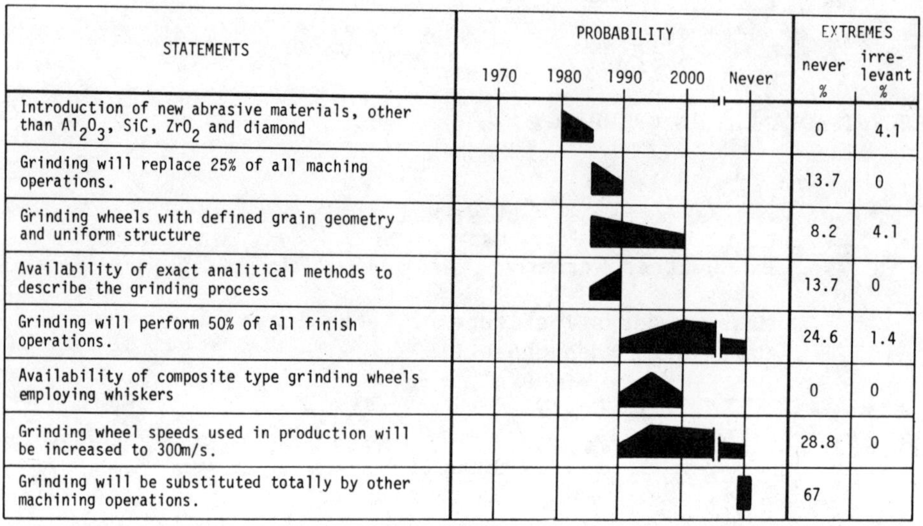

| STATEMENTS | PROBABILITY | | | | | EXTREMES | |
|---|---|---|---|---|---|---|---|
| | 1970 | 1980 | 1990 | 2000 | Never | never % | irrelevant % |
| Introduction of new abrasive materials, other than $Al_2O_3$, SiC, $ZrO_2$ and diamond | | | ■ | | | 0 | 4.1 |
| Grinding will replace 25% of all maching operations. | | | ■ | | | 13.7 | 0 |
| Grinding wheels with defined grain geometry and uniform structure | | | ■ | | | 8.2 | 4.1 |
| Availability of exact analitical methods to describe the grinding process | | | ■ | | | 13.7 | 0 |
| Grinding will perform 50% of all finish operations. | | | | ■ | | 24.6 | 1.4 |
| Availability of composite type grinding wheels employing whiskers | | | | ■ | | 0 | 0 |
| Grinding wheel speeds used in production will be increased to 300m/s. | | | | ■ | | 28.8 | 0 |
| Grinding will be substituted totally by other machining operations. | | | | | ■ | 67 | |

Figure 1: Delphi-Type Forecast of the Future of Grinding (1)

Comparing these statements with today's state of art we realize, that the actual developments are pretty well following the predicted path: New analytical methods are in use to describe and control the grinding processes, and cubic boron nitride is applied as a new synthetic abrasive material for grinding hardened steels and superalloys. Actually the development of grinding technology as achieved in the last ten years can be split into three sections:

- Analysis of technological fundamentals,

- Development of processes and machines,

- Developments of tools and auxiliaries.

In the following chapter a detailed list of the most relevant advances in these three areas is presented together with pertinent references. Subsequently selected topics are discussed in detail.

# RECENT DEVELOPMENTS IN GRINDING

The most important prerequisite for a substantial progress in grinding is the better understanding of the process fundamentals. Consequently, in the recent years major efforts have been directed towards the analytical investigation of the kinematic and mechanical principles of material removal in grinding. At the same time processes have been studied and further developed aiming to achieve higher productivity and increased quality by means of modified ore improved machines, tools and auxiliaries. The following list gives a representative overview on the most important advances in the different areas of grinding technology:

1. Analysis of Technological Fundamentals

1.1 Functional Description of Kinematic Relations (3)

- Number and distribution of static cutting edges,
- Number and distribution of dynamic cutting edges,
- Average chip cross section and chip thickness,

1.2 Functional Description of Grinding Forces (3,4,5,6,7,8)

- Mechanic principles of chip formation,
- Total normal and tangential grinding forces,
- Mean force per individual cutting edge,
- Time-dependent character of grinding force,

1.3 Functional Description of Wheel Wear (2,3,9,10,11,13)

- Mechanic principles of wear at bonded abrasives,
- Difference between radius and edge wear on grinding wheels,
- Cost-optimal grinding conditions,

1.4 Functional Description of Grinding Temperatures (4,12,14,15)

- Basic influence of friction and chip formation,
- State of temperature in contact zone,
- Temperature field in work surface layer,
- Wheel surface temperature,

1.5 Functional Description of Work Surface Roughness (11,16,17)

- Basic roughness due to wheel topography,
- Kinematic roughness due to process conditions,
- Mechanical roughness due to plastic deformations,

2. Development of Processes and Machines

2.1 Advances in Grinding Machine Design (18,19,20,21,22,23,52)

- Application of axiomatic principles for machine tool design,
- Increased static and dynamic stiffness.
- Increased working power,

- Infinitely variable drives for wheel and work,
- Electro-mechanical high-precision drives for tables and slides,
- Improved systems for cooling fluids, application and handling,
- Improved safety measures,
- Improved systems for precision feeding and stopping,
- Improved systems for automatic wheel balancing,
- Integrated and improved dressing devices,
- Temperature control of machine and coolant,

2.2 Advances in Control Techniques (23,24,25,26,27,28,29,30)

- Automatic compensation of wheel wear and dressing losses,
- Automatic control of dressing cycle,
- Control of thermally and mechanically induced distortions,
- Minimizing of idle operation times,
- Optimizing of step grinding,
- Programmable control techniques,
- CNC-control of surface and external plunge grinding,
- Adaptive control techniques,
- Sensor development for force, wear and roughness,
- Establishment of grinding data banks,

2.3 Process Modifications (49,50,51,52,53,54,57,58,59)

- High speed grinding,
- Creep feed grinding,
- High speed creep feed grinding,
- Speed stroke surface grinding,
- Form grinding of worms and gears,
- High speed belt grinding,
- High efficiency thread grinding,
- High efficiency roller grinding,

3. Development of Tools and Auxiliaries

3.1 New Abrasive Materials (31,32,33,34,35,36,38)

- Clustered diamond powder,
- Cubic boron nitride (CBN),
- Zirconia alumina,

3.2 New Bonding Systems (34,37,39,40,41)

- Vitrified bonding for CBN-wheels,
- Brittle bronce bonding for crushable CBN and diamond wheels,
- Resin bonding systems with metal filler and solid lubricants for CBN and diamond wheels,

3.3 New or Improved Dressing Systems (21,42,43,44,45,46)

- High precision diamond rollers,

- Roll-2-dress device,
- Crush dressing of metal-bonded CBN and diamond wheels,
- Dia-dress system for vitrified CBN wheels,
- Single-point dressing for conventional wheels with polycristalline sintered diamond tool marked by positive rake angle,

3.4 New or Improved Auxiliaries (22,24,47,48)

- New clamping system for high speed grinding wheels,
- Grindo-Sonic device for wheel grading,
- Hydraulic wheel balancing system,

The analytical study and description of the grinding process is the key to a better understanding and control. Controversal findings as experienced in using super-abrasives and applying high speed and creep feed techniques can well be resolved in the light of better understood fundamentals. For advanced control techniques, on the other hand, a reliable modeling system, in form of mathematical functions as well as pertinent sensors, is the most important prerequisites for the crucial identification of the controlled processes. Based on these fundamentals and on empirical investigations, the realization of significantly improved productivity and work piece quality in grinding can be achieved by improved machines, tools and auxiliaries. The great variety of recent advances proves, that grinding indeed offers a considerable contribution to increased manufacturing productivity.

SELECTED EXAMPLES OF PRACTICAL ADVANCES IN GRINDING

I. Analytical Description of Grinding Forces

The force generated during grinding is one of the most relevant process criteria because it determines energy consumption, wheel wear and thermal load in the work surface. Using the symbols $v_s$ = wheel speed, $v_w$ = work speed, a = depth of cut, $D = d_w \cdot d_s/(d_w \pm d_s)$ = equivalent wheel diameter, $d_s$ = wheel diameter, $d_w$ = work diameter, $C_1$ = cutting edge density at wheel periphery, $l_k = (a \cdot D)^{1/2}$ = contact length between wheel and work piece, $l$ = variable of contact length, $\overline{Q}(l)$ - average chip cross section as a function of the contact length variable, $N(l)$ = number of active cutting edge as a function of the contact length variable, the following definition for the grinding force per unit of grinding width $b_s$ can be determined from the basic law of chip formation forces (6):

$$F'_n = k \int_0^{l_k} [\overline{Q}(l)]^n \cdot N(l) \, dl \qquad (1)$$

Inserting the functional equations for $\bar{Q}(1)$ and $N(1)$ as derived and described in (2,6), this function yields to a simply structured grinding force model:

$$F'_n = \frac{K}{\varepsilon} \left[C_1\right]^\gamma \left[\frac{v_w}{v_s}\right]^{2\varepsilon-1} \left[a\right]^\varepsilon \left[D\right]^{1-\varepsilon} \quad (2)$$

where: $\varepsilon = \frac{1}{2}\left[(1+n)+\alpha(1-n)\right]$ ;  $0.5 < \varepsilon < 1.0$

$\gamma = \beta(1-n)$ ;  $0 < \gamma < 1.0$ \quad (3)

The exponential coefficient n depends on the material ground and represents the degressive nature of increasing cutting forces versus chip cross section. Its numerical value is always positive and smaller than 1.0. The coefficient $\alpha$ and $\beta$ represent the nature of cutting edge distribution in the wheel periphery and depend on wheel structure, dressing process and state of wear. The proportionality factor K is dependent on the cutting edge geometry, the cooling fluid properties, and the work material properties. Its dimension $kp/mm^2$ (pounds per square inch) is identical to the specific energy dimension.

Equation (2,3) can be derived on a much simpler way, too. Obviously the total grinding force is composed of two different components: friction and chip formation forces. Frictional forces can be related to the actual contact length $l_k = (a \cdot D)^{1/2}$ and the cutting edge density $C_1$. Formational forces, on the other hand, are directly related to the total sum of all instantaneous chip cross section in the contact zone $Q_{mom} = a \cdot v_w / v_s$. Superimposing these two terms such that the influence of both varies reciprocally between 0 and 1 by employing an exponential coefficient $0 < \mathcal{S} < 1$:

$$F'_n = C_1^\gamma \left[K\sqrt{a \cdot D}\right]^{\mathcal{S}} \cdot \left[K \cdot (a \cdot v_w / v_s)\right]^{1-\mathcal{S}} \quad (4)$$

$\underbrace{\qquad\qquad\qquad}_{\text{frictional term}}$ $\underbrace{\qquad\qquad\qquad}_{\text{formational term}}$

leads to exactly the same equation for the grinding force as (2,3) by applying the substitution $\mathcal{S} = 2 - 2\varepsilon$. Thus, the nature of grinding forces as a superimposed system of frictional and formational forces, which both depend on mechanical, kinematic and geometric process parameters, has been proven.

For practical applications it is important to notice, that materials with high ε-values near 1.0 have a very good grindability with respect to thermal conditions. They have a ductile-hard character and don't generate much friction in micro-cutting; they form regular chips, do not tend to load the wheel surface with work material, and show a clear tendency for constant work temperatures when the wheel speed is increased. Materials with low ε-values near 0.5 have a rigid-hard or ductile-soft nature, they are marked by a very high component of frictional forces in chip formation, form a debris-type of chip, and show a significant tendency for increased work temperatures when the wheel speed is increased. These material-dependent characteristics of the micro-chip formation process in grinding are being studied extensively in Europe, and grinding data banks are being established at the same time,

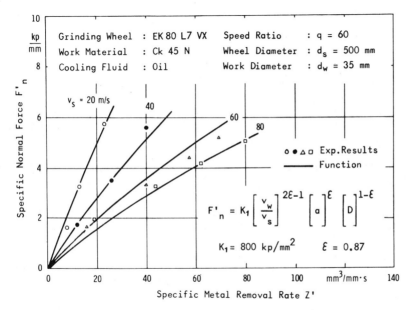

Figure 2: Grinding Force Model Applied to External Plunge Grinding Results

thus, making use of the analytical and practical results of university-industry cooperation (24,5). As an example for these activities Figure 2 shows the application of the described force model to practical results obtained in external plunge grinding of a medium carbon steel (equivalent to AISI 1045) being ground with a conventional $Al_2O_3$ wheel using oil as coolant.

II. Application of Higher Wheel Speeds and Temperatures Aspects

The energy consumed in grinding per unit of grinding width $b_s$ can be understood as the product of the tangential grinding force $F'_t$ and the wheel speed $v_s$. Assuming as a first approximation, that the amount of energy flowing into the work surface is a constant proportion of the total energy, whereas the rest becomes effective in the chips, the grinding wheel and the cooling fluid, then the maximum temperature $T_{max}$ in the work surface can be regarded as proportional to the total energy E per unit of grinding width:

$$T_{max} \approx E = F'_t \cdot v_s \tag{5}$$

This means that the work surface temperature increases proportionally with the wheel speed $v_s$, provided that the tangential grinding Force $F'_t$ per unit of grinding width remains fairly constant versus $v_s$. This is exactly true when the coefficient $\varepsilon$ in (2,3) is near 0.5, i.e., with brittle-hard or ductile-soft materials. However, when $\varepsilon$ has a value near to 1.0, as valid for ductile-hard materials with good grindability, then the tangential force decreases at the same rate the wheel speed goes up, and consequently the work surface temperature must remain constant and independent from the wheel speed $v_s$ in this case.

This interrelation has clearly been proven by practical investigations (56), and it is well known by general practical experience, too. <u>Figure 3</u> shows the temperature behavior of eight different materials, obtained by practical measurements in surface grinding. Basically two groups of materials can be distinguished: The first one does on an average not show higher work surface temperatures when the wheel speed is increased 100% from 30 m/s (6000 sfpm) to 60 m/s (12000 sfpm). This group consists of the following materials: High-alloyed steel X 210 Cr 12 (Chromium steel with Cr = 12%, C = 2.1%), high speed steel S 6-5-2 (equivalent to Mi steel), medium carbon steel Ck 45 N (equivalent to AISI 1045), and cobalt-base superalloy ATS 115 with Co = 52%, Cr - 20%, W = 15%, Ni = 11%, and all of them do qualify for the application of increased wheel speeds. The other group shows a significant increase of work surface temperature when the wheel speed is doubled: Low carbon steel C 15 (equivalent to AISI 1015), cast iron with globular graphite GGG 70, nickel-base superalloy RGT 12 with Ni = 58%, Cr = 20%, Co = 18%, Ti = 2%, Fe = 2%, and austenitic stainless steel Remanit 1880 SST with Cr = 17%, Ni = 12%, Mo = Ti = 2%. These materials are not suited for grinding with higher wheel speeds, because they are marked by a high frictional force component indicated by a low $\varepsilon$-value. Consequently, this second group of materials shows a more significant drop of temperature when oil instead of emulsion is used as coolant, thus, reducing the considerable frictional energy by better lubrication.

If these simple force-related fundamentals would have been known ten years earlier, a lot of frustrating attempts could have

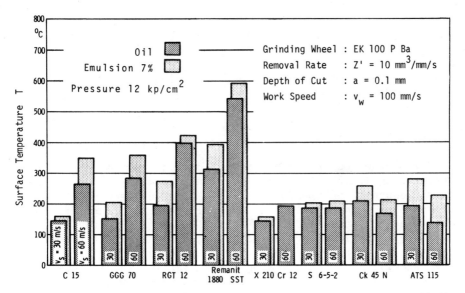

Figure 3: Surface Temperatures at Increased Wheel Speeds (56)

been avoided, aiming to increase productivity in grinding at all means by applying higher wheel speed with all materials. In the bearing industry where group-1 type of materials with outstanding grindability are used by tradition, high speed grinding is regarded as a strong and successful method to increase productivity. In the turbine blade industry materials of group-2 type are used mainly; here the implementation of high speed grinding was bound to fail.

Some examples for successful application of increased wheel speeds in production grinding are:

a) External cylindrical grinding of standard ball bearing rings (20). The race of the internal ring is ground with 120 m/s (60000 sfpm) on a FAMIR RTF-C2 high speed grinding automat, using a vitrified sandwich-type $Al_2O_3$ wheel and an emulsion as coolant. The part time was reduced from 39 to 2.1 seconds, a 2000% productivity gain resulted in the production of 1370 races per hour, and production cost could be reduced 75%.

b) High speed - creep feed grinding of twist drill flutes made from S 6-5-2 high speed steel, which is according to Figure 3 well suited for higher grinding wheel speeds (51). Both flutes are ground out of the annealed full material in one pass on a specially designed machine, using resin bond $Al_2O_3$ wheels at 120 m/s (60000 sfpm) and oil as coolant.

Achieved metal removal rates of 115 $mm^3/mm/s$ are 15 times higher than in milling the flutes, and surface integrity standard is clearly increased.

c) Crank shaft grinding with wheel speeds increased up to 80 m/s (40000 sfpm) and increased metal removal rates (57). Work material: chilled iron, hardness: $HR_c = 50$.

## III. Creep Feed Grinding

One of the most promising recent developments in precision manufacturing is the creep feed grinding method. In comparison with the conventional pendulum surface grinding technique this process provides a great potential to increase productivity and accuracy at the same time (19). This is especially true if deep and/or profiled slots have to be machined in difficult to machine work materials. Furthermore, creep feed technology offers an improved stability of the grinding wheel profile, and it shows a considerable reduction of thermal effects in the work surface.

Generally the creep feed surface grinding process is marked by a specific mode of operation. In contrast to the conventional pendulum technique, the depth of cut is increased 100 to 10000 times and the work speed is decreased in the same proportion. Thus, it is possible to grind slots with a depth of 1.0 to 30.0 mm (0.04 to 1.2") and more in one pass, using work speeds from 0.75 to 0.025 m/min (30 to 1 ipm), and reducing machining times 40 to 80%. In addition, profile stability is increased considerably.

An important prerequisite for making full usage of the economic and technological advantages of this high efficiency/high precision technology is the application of specially developed and constructed machine tools, grinding wheels, dressing methods, and controlling techniques. Such a system should provide the following features:

- High static and dynamic stability of machine tool,
- High accuracy, stick-slip free slides with favorable damping characteristics,
- Considerably increased spindle power,
- Infinitely variable spindle revolution number,
- High-balanced and directly connected motor-spindle system,
- Pre-tensioned, high performance, high accuracy spindle bearings
- Non-hydraulic, single unit table drive covering the whole work speed area from creep feed to

pendulum region,
- High pressure cooling and cleaning system,
- Interated dressing devices,
- Pertinent grinding wheels,
- Up-dated process know-how.

Conventional surface grinders cannot at all provide these advanced features, and even modified conventional machines cannot meet with most of these necessary requirements. This is the reason why many attempts to change over to creep feed technology failed. The first creep feed prototype for surface grinding was developed in 1958, and in 1963 the first couple of production creep feed grinders were applied in practice (58). Since then this technology has steadily been improved, and application has grown significantly. Today several grinding machine manufacturers offer creep feed machines, even for external plunge grinding, and as a consequence this technology can now be regarded as a well established and competitive manufacturing alternative.

Figure 4 illustrates the major technological differences between conventional pendulum and creep feed grinding. Total grinding forces, single edge cutting forces and maximum surface temperatures are plotted versus depth of cut and work speed. The specific removal rate Z' is the product of both the depth of cut and the work speed ($Z' = a \cdot v_w$), and it is kept constant. Thus, the left part of the diagrams represent pendulum conditions, whereas the right parts refer to creep feed conditions.

It is obvious, that the total grinding forces increase degressively with greater depth of cut. As a consequence, in creep feed grinding the power requirements and the machine stiffness must be significantly higher compared with conventional surface grinding machines. The reason for this specific characteristic is the much larger contact length between wheel and work piece at creep feed conditions. In strong contrast to this, the mean force per individual cutting edge decreases significantly versus depth of cut, and is much lower at creep feed conditions. As a consequence, wheels used for creep feed operations must show a reduced hardness, because otherwise the wheels tend to a glazing type of wear. In other words, the average load per cutting edge would not be high enough to secure the required steady state of wear. In practice this phenomenon is well recognized, however, the very reason why softer grinding wheels are necessary in creep feeding is not really understood. Mostly it is explained by the influence of a more open wheel structure as applied in creep feed grinding because more cooling fluid can be carried into and through the contact zone by such a wheel.

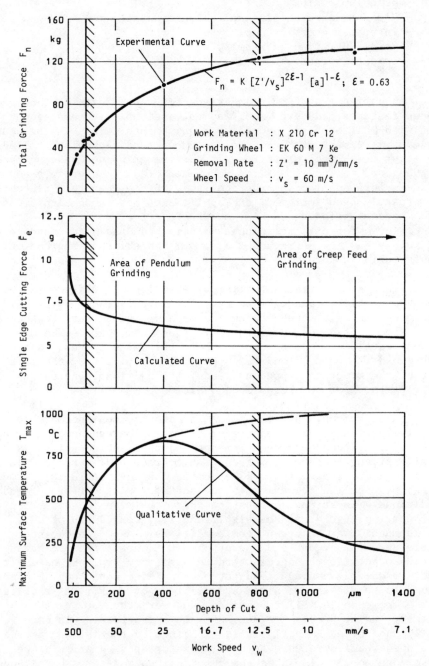

Figure 4: Behavior of Grinding Forces and Work Surface Temperatures in Creep Feed and Pendulum Regimes (19)

The most important prerequisite for the successful application of creep feed technique as an advanced manufacturing method is the very low thermal load as being induced into the surface layer of the work piece. Figure 4 shows, that the maximum surface temperature increases first with greater depth of cut, then it reaches a maximum, and towards the creep feed region decreases steadily. This observation seems to be in contradiction to the permanently increasing grinding force, indicating that the total energy consumed increases permanently, too versus depth of cut. However, with correspondingly lower work speeds the increasing amount of thermal energy flowing into the work surface, has more time to dissipate deeper into the work material. Thus, a larger amount of energy is effective in an even larger work volume, and results in reduced temperatures. If this specific mechanism would not be in effect then creep feed grinding would not be possible at all.

Figure 5 shows a modern surface Grinder specially designed for creep feed and pendulum operations (58). This model has all features and characteristics necessary for creep feed operations as mentioned above. The work piece illustrated in Figure 6 is a typical example for creep feed application. Turbine blades like this made from nickel-cobalt-based superalloys are ground in one or two roughing passes and a subsequent finishing pass. Frequently both the profiles are cut at the same time by means of a double-disk creep feed grinder. The grinding wheels are dressed and profiled by high precision diamond roller dressing devices (44). With work speeds less than 0.1 m/min (4 ipm) and oil or emulsion as coolants, a very good surface quality in terms of roughness, thermal integrity, and profile accuracy are accomplished. In addition the floor to floor times are reduced 25 to 50%. - Another frequent example is the slotting of air pressure rotors and hydraulic pump rotors. In these cases milling and subsequent finish grinding was often substituted by one single creep feed pass out of the full and often hardened material. Savings of machining time do run up to 75% especially when long slots with small widths are given.

One important constraint for creep feed grinding is its sensitivity with regard to deviations from the optimal working conditions. Because of this, creep feed technology can economically be applied only, when at mass production of suited parts the special know-how is generated and conserved by a group of dedicated engineers and operators, provided that the best equipment in form of machines, grinding wheels, dressing tools and fixtures are available.

IV. Speed Stroke Grinding

Speed stroke grinding is a new surface grinding technique, which employs a normal pendulum surface grinder with a newly

Figure 5: ELB Creep Feed Grinder, Type W08 (58)

Figure 6: Turbine Blade Ground by Creep Feeding

developed table drive system: A thyristor-controlled DC motor
moves the work table back and forwards by means of a pre-tensioned
gear belt. This system has a significantly higher degree of control, and as the crucial feature it provides a much higher table
stroke rate at small stroke lengths. Thus it overcomes one of the
major deficiencies of conventional hydraulic table drives, as they
are in common use today (59).

Figure 7 clearly demonstrates the superiority of the new
method over the old technique: At a stroke length of $L_s<50$ mm
(2") a conventional hydraulic drive produces 30 strokes per minute. With the new speed stroke system up to 300 strokes per minute, that are 5 strikes per second, are achieved, and productivity
is increased 10 times. Actually, the table speed is kept in the
economic range above 10 m/min (30 fpm). With greater stroke length,
i.e., larger work pieces, the differences in stroke rates are less
pronounced, however, there is still a 30% advantage in this regime
for speed stroke grinding to be used. As a second advantageous
feature the grinding wheel is not fed stepwise after each pass

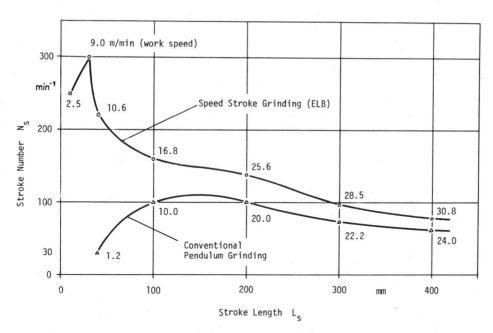

Figure 7: Stroke Rate Versus Stroke Length for Speed Stroke and
Conventional Conditions (59)

down into the work piece but moves in continuously. This infeed is
automatically controlled in relation to the stroke rate such that
the grinding wheel never comes out of contact with the work piece
until the operation is finished. Thus, the contact shocks as effective in normal pendulum grinding at the beginning of each individual stroke, are eliminated in speed stroke grinding, resulting
in a significantly reduced wheel wear.

Speed stroke grinding is best suited for grinding slots and
profiles in small work pieces, and when the number of parts to be
ground is small, as given normally in shop and tool room grinding.
However, also for small and medium batch production of small work
pieces it could become an economic alternative to creep feed grinding, because the machine is significantly less expensive, the technological know-how required with regard to grinding wheels and cooling fluids is rather conventional, the metal removal rates are
equivalent or even better, and the speed stroke system needs less
service and set-up time, than creep feed machines.

V.  Cubic Boron Nitride (CBN) a New Abrasive Material

Cubic boron nitride is a crystal synthesized by a high-temperature/high-pressure treatment. Simular to the graphite-diamond transformation the hexagonal boron-nitride powder is thereby changed into the cubic CBN crystal. CBN is commercially available since 1969 as a GE product (55), and was first used as abrasive
material in form of resin bounded wheels for grinding of high speed
and tool steels. More and more, however, CBN conquers the area of
production grinding of superalloys, hardened steels of all kinds,
and chilled cast iron. It has properties which are somehow similar
to diamond, but some decisive minor differences (Table 1) result
in completely different fields of applications for both of these
so-called "superabrasives."

Diamond is by far the hardest material of all. Therefore, it
is predestinated to grind rigid-hard materials like sintered carbides, refractory materials, and granites. However, diamond is
sensitive against elevated temperatures, especially in the presence
of oxygen as always effective in normal grinding processes. In
grinding the above mentioned rigid-hard materials the critical oxidation temperature of about $700^o C$ is not reached. With ductile
materials like soft or hardened steels, however, the effective chip
formation temperatures are in the range of $1200^o C$ and more. As a
consequence, the hardest abrasive is not qualified to grind steels.
CBN, however, has a much higher oxidation temperature and in spite
of its lower hardness it is well suited to grind steels especially
in the hardened state. Due to the lower hardness CBN on the other
hand cannot effectively grind sintered carbides and other
refractories.

In order to increase the application of CBN in production grinding two major obstacles must be overcome. The first one is a non-engineering problem: the extreme high price of CBN. With increasing usage there will be a good chance that prices go down. Competition, if encouraged, could help, too. The second problem is of mechanical nature: it is extremely difficult to dress and true CBN wheels to the required shapes and sharpness, especially when profiled wheels are used.

Table 1: Hardness and Thermal Properties of Different Abrasives

| Property | CBN | Diamond | $Al_2O_3$ |
|---|---|---|---|
| Hardness $[kp/mm^2]$ | 4200 | 9000 | 2100 |
| Decomposition Temperature $[°C]$ | 1550 | 1400 | 2400 |
| Oxidation Temperature $[°C]$ | 1300 | 700 | -- |

The main problem still is, that after dressing with diamond dressing devices CBN wheels do not cut because the dressed surfaces of common resin and metal bonded wheels are completely closed and glazed. Only after a lengthy truing process is applied by treating the surface with a sintered $Al_2O_3$ stick for 30 to 60 minutes and more, depending on the wheel size, the BZN wheel might become sharp. This is an absolute impractical method, which is rejected unanimously in industry.

Some recent developments seem to clear the way out of these difficulties. The first one is a metal bond of brittle character (44) enabling to profile CBN and diamond grinding wheels by means of the well known crushing method. A roller from hardened steel, sintered carbide, or though refractories, which possess the negative wheel profile, are forced into the revolving wheel, thus, working the profile into its periphery. Because of the very high grinding ratios (G = work material removed/worn wheel volume) those tools achieve, the loss of CBN grains during crush dressing can well be afforded. Figure 8 shows a crushing tool penetrating into a brittle bronce CBN wheel. Recent practical results, especially in creep feed grinding are extremely positive with regard to simplified dressing, achieved removal rates and grinding ratios, as well as the resulting profile accuracy and life time. Due to special structural components, these brittle-bronce wheels show a significantly lower grinding temperature than usually experienced with metal bonded grinding wheels (44).

Figure 8: Crush Dressing of a Brittle Bronce CBN Grinding Wheel (44)

The second novelty are vitreous bonds for internal grinding CBN wheels. Vitreous bonds are expected to provide a much better dressability, however, the conventional type requires a baking temperature near the decomposition temperature of CBN. This problem has been solved now by the development of a new glass-type of bond with significantly reduced processing temperature (45). Figure 9 shows the dressed surface of such a CBN wheel. It has an

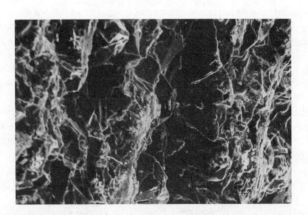

Figure 9: Surface of Vitreous Bond CBN Wheel for Internal Grinding Operations (37)

open appearance and cuts very free and cool (37). As an additional advantage, the vitreous bond acts as an abrasive component, too, resulting in higher performance, controlled bond wear, and exceptionally good surface roughness. Figure 10 represents some results of recent investigations achieved at the Technical University in Hanover, West Germany (39). In internal grinding of hardened ball bearing steel these wheels showed extremely high grinding ratios of G = 4000 to 16000, depending on the abrasive grit size and the applied metal removal rate. Obviously the system is sensitive against mechanical overloading, because the tests reveal a strong decreasing gradient of the G-ratio with higher removal rates, especially in the case of the finer grit size. Therefore, these tools need a careful selection and control of the applied working parameters. During the tests all other conditions have been kept identical to conventional grinding with $Al_2O_3$ wheels. For dressing a specially designed, fast rotating, electro-plated cup-type of diamond tool has been used (45).

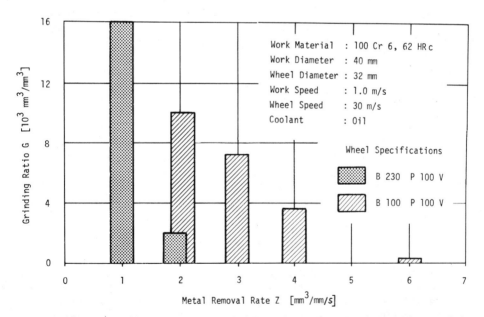

Figure 10: Grinding Ratios as Achieved in Internal Grinding with Vitreous Bond CBN Wheels (39)

It is very likely, that developments like these will in the near future lead to a highly increased consumption of CBN abrasives in production grinding. Other areas of potential applications are gear grinding, thread grinding and centerless grinding.

VI. New Sensor for In-Process Detection of Work Surface Roughness

The work surface roughness is, besides the thermal integrity, the most important criterion of the working result in grinding. The only way, however, to check on surface roughness in production is to measure it externally by means of a conventional surface analyzer, which use a stylus to pick up a two-dimensional representation of the three-dimensional asperity distribution of the work surface. This on principle is a very slow process, because during measuring the effective profile frequency must be clearly below the natural frequency of the pick-up system to avoid overriding.

A newly developed detection method turns things round and makes beneficial use of this disturbance by having a stylus responding in his natural frequency to the high and random frequency of a fast moving work surface. As a result the average stylus amplitude comes out as being proportional to the average asperity height of the work surface, and to the work speed (27).

Figure 11 shows the system in principle. In the diagram at the right hand side the total roughness R and the center line average $R_a$ as valid for different work surfaces are plotted against the recorded system out-put S. The constant work surface speed of 50 m/min is at the higher end of the range for conventional speeds in external grinding, and it is more than 1000 times higher than applied with conventional roughness analyzers.

The new system was originally developed to work as an in-process sensor for an adaptive control (AC) external grinding system. It might however, be of more immediate use as a module for in-process control of surface roughness in conventional external and centerless grinding operations. First practical experiences in production are being made actually at a major German automobile manufacturer.

VII. Electro-Sonic Determination of Grinding Wheel Hardness

Measuring the hardness of grinding wheels is an important operational process in wheel production and application. However, the traditionally used methods were inaccurate and destructive to a certain degree. An example for this is the sand blasting method where sand is blasted at defined rates, pressure and time onto the

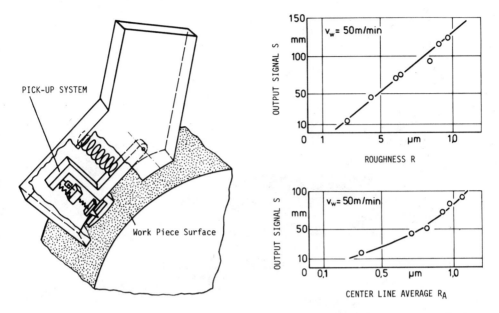

Figure 11. Principles of New Method for Rougness Detection (27)

wheel surface, and the depth of the created hole is taken as a reference for the wheel hardness or softness respectively.

Since 1970 a new non-destructive method, the Grindo-Sonic system (60) is available (Figure 12), by means of which the hardness is determined from the elastic characteristics of the entire wheel volume. Actually, the grinding wheel or any other body like honing stones, sintered carbide parts or composite structures, are positioned such that they can vibrate free. Then after a slight impact with a screw-driver handle or a resin stick, the natural vibrations are picked up by a microphone and processed in a specially designed micro-loging unit. Taking the specific geometrical conditions into account, the hardness of the investigated body can be read from the display immediately. Some of the effective advantages are:

- 100% increased accuracy in hardness determination,
- Investigation on uniformity of structures possible by impacting different places,
- Same accuracy with large and small parts,
- Unlimited repeatability,

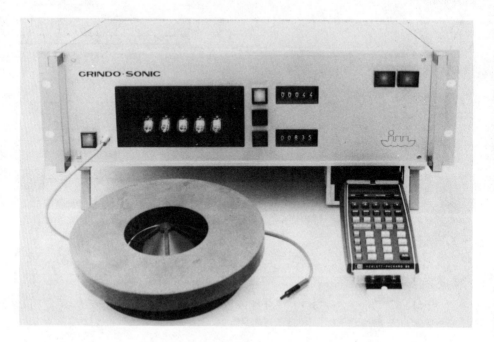

Figure 12: Grindo-Sonic Instrument for Wheel Hardness Testing (60)

- Predominant consideration of the external structural volume, which are most relevant for the working hardness of grinding wheels,
- 100% checks are possible,
- low operational cost.

Today some 200 Grindo-Sonic instruments are applied in practice all over the world. They are not only used by manufacturers and users of abrasive tools, but also in the ceramic and glass industry, graphite manufacturers, construction industry, sintered metal and carbide manufacturers, and in manufacturing laboratories.

## References

1. Merchant, M. E., Delphi-Type Forecast of the Future of Production Engineering, CIRP-Reports, Vol. 20/1, 1971.

2. Peklenik, J., Determination of Geometric and Physical Parameters for Basic Research in Grinding (German), Dissertation, Techn. University Aachen, 1957.

3. Werner, G., Kinematics and Mechanics of the Grinding Process (German), Dissertation, Techn. University of Aachen, 1971.

4. Werner, G., Koenig, W., Adaptive Control Optimization of High Efficiency External Grinding - Concept, Technological Basics and Application, CIRP Annals, Vol. 23/1, 1974.

5. Decneut, A., Basic Description of the Grinding Conditions (Flemish), Dissertation, University Leuven, 1974.

6. Werner, G., Influences of Work Material on Grinding Forces, CIRP Annals, Vol. 27/1, 1978.

7. Werner, G., Concept and Technological Fundamentals of Adaptive Control Optimization in External Grinding (German), Habilitation, Techn. University Aachen, 1973.

8. Saljé, E., New Results on the Time-Dependency of Grinding (German), Technische Mitteilungen, No. 7/8, 1976.

9. Werner, G., Relation Between Grinding Work and Wheel Wear in Plunge Grinding, SME-Technical Paper, MR 75-610, 1975.

10. Weinert, K., Time-Dependent Alteration of Wheel Periphery in External Grinding (German), Dissertation, University Braunschweig, 1976.

11. Bierlich, R., Technological Prerequisites for Adaptive Control Systems in Plunge Grinding (German), Dissertation, Technical University Aachen, 1976.

12. Malkin, S., Anderson, R. B., Thermal Aspects of Grinding, Transactions of ASME, Nov. 1974.

13. Malkin, S., The Attritious Wear and Fracture Wear of Grinding Wheels, Dissertation, Mass. Institute of Technology, 1968.

14. Sauer, W. J., Thermal Aspects of Grinding, Dissertation, Carnegie Mellon University, Pittsburgh, 1971.

15. Snoeys, R., Maris, M., Peters, J., Thermally Induced Damage in Grinding, CIRP Annals, Vol. 27/2, 1978.

16. Fruehling, R., Topography of Wheel Periphery and Surface Roughness in Plunge Grinding (German), Dissertation, Techn. University Braunschweig, 1976.

17. Werner, G., Functional Description of Work Surface Roughness in Grinding (unpublished), Research Project, Mass. Institute of Technology 1978.

18. Suh, N. P., Bell, A. C., Gossard, D. C., On an Axiomatic Approach to Manufacturing and Manufacturing Systems, Transactions of ASME, 1978.

19. Werner, G., Applications and Technological Fundamentals of Creep Feed Surface Grinding, Unpublished Investigation, Mass. Institute of Technology, 1978.

20. Bogusch, E., Leichter, J., High Speed Grinding up to Wheel Speeds of 120 m/s (German), Jahrbuch der Schleiftechnik, No. 47, Vulkan-Verlag, Essen, 1977.

21. Meyer, H. R., Dressing with Diamond Roller Dressers (German), Werkstattstechnik, No. 65, 1975.

22. Lauer-Schmaltz, H., Practical Experiences with a New Automatic Wheel Balancing Device (German), Jahrbuch der Schleiftechnik, No. 48, Vulkan-Verlag, Essen, 1978.

23. Mohr, H., Numerical Controlled Flat and Cylindrical Profile Grinding, (unpublished seminar presentation), ELB-Grinders Corporation, Mountainside, N.J., 1977.

24. Koenig, W., and others, Staying Competitive by Using Technological Reserves (German), Presentation at 16th Machine Tool Colloquium, Techn. University Aachen, 1978.

25. Saljé, E., Grinding Processes Considered as Feed Back Systems, CIRP Annals, Vol. 27/1, 1978.

26. Piepenbrink, R., Programmable Control for Internal Grinding, Jahrbuch der Schleiftechnik, No. 48, Vulkan-Verlag, Essen, 1978.

27. Scherf, E., Mushardt, H., Roughness Sensor for Optimizing of Grinding (German), Jahrbuch der Schleiftechnik, No. 48, Vulkan-Verlag, Essen, 1978.

28. Sata, T., and others, In-Process Measurement of Grinding, Preceedings of the International Grinding Conference, Carnegie-Mellon University, Pittsburgh, 1972.

29. Volk, J. F., Chase, R. P., Adaptive Controls, Machinery, December 1972.

30. Werner, G., Advanced Concepts of Wear-Cost-Related Optimization in External Plunge Grinding, SME-Technical Paper, MF 76-156, 1976.

31. Buettner, A., Grinding of Rigid-Hard Materials with Diamond Cup Wheels with Special Regard to Creep Feed Operations (German), Dissertation, Techn. University Hanover, 1968.

32. Lindsay, P., Navarro, P., Principles of Grinding with Borazon CBN Wheels, Machinery, May/June 1973.

33. Mishnaevski, L., Use of Cubic Boron Nitride Wheels in Gear Grinding, Machines and Tools, USSR, Vol. XLIV, No. 2.

34. Meyer, H. R., Economic Application of CBN and Diamond Wheels for Cylindrical Plunge and Profile Grinding (German), Technische Mitteilungen, No. 6/7, 1976.

35. Kaiser, M., Pendulum and Creep Feed Grinding of Sintered Carbide with Diamond Wheels (German), Dissertation, Techn. University Hanover, 1975.

36. Pung, R. C., Centerless Grinding with Borazon CBN - A Status Report, Seminar of Industrial Diamond Association of Japan, Tokyo, 1978.

37. Toenshoff, H. K., Juergenhake, B., Superhard Abrasives and Matching Bonds for Internal Grinding, CIRP Annals, Vol. 27/1, 78.

38. Jakobs, U., State of Art in Grinding with Cubic Boron Nitride Wheels (German), Jahrbuch der Schleiftechnik, Vulkan-Verlag, Essen, No. 48, 1978.

39. Juergenhake, B., Cost-Advantages in Internal Grinding Using Vitrified CBN Wheels (German), ZWF - Zeitschrift fuer wirtschaftliche Fertigung, No. 10, 1978.

40. Renard, P. A., Smith, L. I., An Innovative Bond System Extends the Utility of CBN Abrasives, Industrial Diamond Review, May 78.

41. Siqui, R. H., Cohen, H. M., Resin-Bonded Abrasive Tools with Metal Fillers, United States Patent No. 3779727, Dec. 1973.

42. Notter, A. T., Bailey, M. M. Trueing and Dressing Diamond and CBN Grinding Wheels, Industrial Diamond Review, May 1977.

43. Sawluk, W., New Technique for Dressing of Diamond and CBN Peripheral Grinding Wheels (German), Technische Mitteilungen, No. 6, 1974.

44. Meyer, H. R., Profile Dressing of Diamond and Cubic Boron Nitride Grinding Wheels (German), Jahrbuch der Schleiftechnik, Vulkan-Verlag, Essen, No. 48, 1978.

45. Hartl, H., Internal Grinding with Borazon CBN Wheels (German), Seminar Presentation, Tyrolit Schleifmittelwerke KG, Schwaz, Austria, 1977.

46. Skinner, F. R., Compax Blanks - A Positive Approach to Dressing, Seminar of Industrial Diamond Assoc. of Japan, Tokyo, 1978.

47. Nowak, E., Development of Clamping Systems for Vitrified Grinding Wheels (German), Dissertation, Techn. University Braunschweig, 1975.

48. Decneut, A., Non-Destructive Hardness Testing of Abrasive Tools (German), Jahrbuch der Schleiftechnik, Vulkan-Verlag, Essen, No. 46, 1976.

49. Hellwig, W., Roller Grinding Machines with Increased Removal Rates (German), Jahrbuch der Schleiftechnik, Vulkan-Verlag, Essen, No. 48, 1978.

50. Moore, K., Putting the Brakes on Speed, American Metal Market - Metalworking News Edition, Grinding and Finishing Section, March 13, 1978.

51. Geuhring, K., On New Prospects in High Speed Grinding (German), Jahrbuch der Schleiftechnik, Vulkan-Verlag, Essen, No. 46, 1976.

52. Stauffer, R. N., Breakthrough in End Mill Grinding, Manufacturing Engineering, July 1978.

53. Wick, C. H., The Advent of Creep Feed Grinding, Manufacturing Engineering and Management, June 1975.

54. Druminski, R., Thread Grinding with Cubic Boron Nitride Wheels (German), Industrie-Anzeiger 99, No. 4, 1977.

55. Devries, R. C., Cubic Boron Nitride - Handbook of Properties, General Electric Company, Corporate Research and Development, Technical Information Series, Report No. 72 CRD 178, June 1972.

56. Koenig, W., Dedrichs, M., Surface Grinding with High Wheel Speeds and Metal Removal Rates, MTDR Conference, Birmingham, Sept. 1972.

57. Vetter, U., Brill, J., Grinding of Non-Circular External Parts (German), Jahrbuch der Schleiftechnik, Vulkan-Verlag, Essen, No. 48, 1978.

58. Lang, E., Herzig, H., Development of Prototype and Machines for Creep Feed Surface Grinding, ELB-GRINDERS Corporation, Mountainside, N.J., 1958-1978.

59. Herzig, H., Unpublished Report on Investigation of Speed Stroke Grinder, ELB-Schliff, Babenhausen, W. Germany, 1978.

60. Lemmens, J. W., Decneut, A., Private Information and Courtesy, J. W. Lemmens - Electronica N. V., Leuven, Belgium, 1978.

MATHEMATICAL AND ECONOMIC MODELS FOR MATERIAL REMOVAL PROCESSES

Vijay A. Tipnis*

Vice President and Director of Manufacturing
Technology
Metcut Research Associates, Inc.
Cincinnati, Ohio

ABSTRACT

Since current phenomenological models of material removal processes do not provide predictive relationships, empirical models (deterministic, statistical, probabilistic and stochastic) have been developed and applied to material removal operations. A methodology for the development and application of these models is presented. Such models are vitally needed for the selection of economic machining conditions within the given constraints of the cutting tool, the machine tool and the workpiece.

Economic models are needed for the creation of cost-effective process plans, especially with the computer-aided process planning systems. Based on a rigorous generalized economic model for material removal process introduced earlier, macro- and micro-economic models have been developed. The macro-economic model is applicable to cost estimation and pre-planning; the micro-economic model is essential to the selection and optimization of operating parameters. The overall organization of the data generation, collection, analysis and implementation of the process and economic models is presented.

NOMENCLATURE

AD, RD = axial and radial depths (in.), respectively
$H_c$, $H_t$ = productivity functions
I, c, t = generalized economic function; cost function ($/piece); and time function (min./piece), respectively
R = process rate function, e.g., cutting rate or cutting rate function for material removal operations (cu.in./min.)

T = process interruption function, e.g., tool life or tool life function for material removal operations (min.)
$t_d$ = duration of interruption, e.g., tool change time (min.)
n = batch size
$t_s$ = setup time for the batch
$t_L$ = load/unload time per piece
$l_r$ = rapid traverse distance
$l_c$ = length of cut
e = extra travel (air feed)
r = rapid traverse rate
V = cutting speed (fpm)
v = volume of air cut at feed (cu.in.)
$v_a$ = volume of air cut at feed (cu.in.)
M = work center rate ($/min.)
Y = tool cost/usage (purchase cost + regrinding cost/no. of regrinds + 1), ($/usage)
z = number of flutes on milling cutter
$\lambda_0$, $m_0$, $k_0$ = generalized constant coefficients; (setup time/piece + load/unload time + rapid traverse time); ($m_0$M), respectively
$\lambda_1$, $m_1$, $k_1$ = generalized process coefficient, (v + $v_a$) and ($m_1$M), respectively
$\lambda_2$, $m_2$, $k_2$ = generalized auxiliary coefficient; (v x $t_d$) and ($m_2$M + Y), respectively
$\lambda_3$, $m_3$, $k_3$ = generalized material expenditures, expenditure of time for material preparation and material cost, respectively

INTRODUCTION

Material removal processes such as machining and grinding lack adequate quantitative characterization. This presents a major impediment to the improvement of process design, planning, optimization and control.

The necessity for transforming the aerospace manufacturing industry from "experience-based" to "knowledge and data-based" has become increasingly evident with the introduction of computer aided manufacturing. Adequate quantitative characterization in terms of usable mathematical models of the specific material removal processes involved is the first crucial step in the transformation.

In this paper, a methodology for the development of empirical models of material removal processes is presented. Also, comprehensive economic models for macro- and micro-economic analysis of material removal processes are presented.

## PHENOMENOLOGICAL APPROACH

The basic characteristics of the chip removal process can be described schematically as shown in Figure 1. Chip removal is achieved through intense shear deformation within the shear zone. Another important feature is the formation of the built-up edge, which is a highly deformed portion of workpiece material adhering to the tip of the cutting tool. Unstable built-up edge is one of the prime causes of poor surface finish. About 75 percent of the energy generated during material removal is consumed in the shear zone, regions 1 and 2, and the remaining energy is consumed at the frictional interfaces, 3 and 4. The shear zone and frictional interfaces characteristically have high values of combined shear and normal stresses. The shear zone involves intense shear deformations at strains as high as 5 in./in./sec. The temperature in the shear zone and at the frictional interfaces may range from about 800°F to 1400°F depending on whether HSS or carbide tools are used.

One of the major difficulties in applying a phenomenological approach to material removal processes has been that the material behavior relevant to a material removal process cannot be reproduced and studied except through actual material removal tests. In other words, meaningful data on material behavior such as stress-strain, strain rate and temperature as they relate to material removal operations cannot be obtained from mechanical testing.

Over the last 90 years of research in this area, several different phenomenological approaches have been tried for the development of usable quantitative characterizations of the material removal processes (see Table I). Among some of the important approaches tried have been: (a) mechanics of chip formation; (b) plasticity analysis using slipline theory, (c) thermal analysis, (d) tool wear theories, and (e) metal physics and dislocation study of the shear zone. Each of these approaches has produced useful information; however, none of these approaches has been able to produce a usable theory or relationships that can be used to predict important machining responses such as tool life, as-machined surface finish and accuracy, given the work material, the cutting tool and the operating conditions.

## EMPIRICAL APPROACH

The recommended approach for quantitative characterization of the material removal process is that of developing mathematical relationships between the machining response and the operating conditions directly through a set of experiments. The approach involves the application of statistically planned experiments and the development of statistically valid, deterministic and proba-

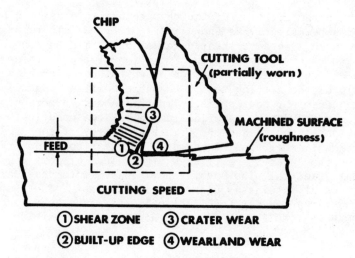

Figure 1. Phenomenological Characteristics of Material Removal Process - Schematic Representation

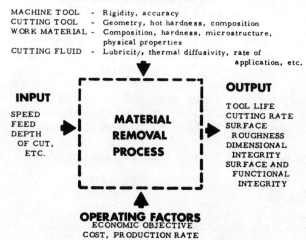

Figure 2. Characteristics of Material Removal Process for Empirical Models

# MODELS FOR MATERIAL REMOVAL PROCESSES

blistic mathematical models of material removal processes. The mathematical relationships developed through this approach are directly applicable to the material removal processes. It should be noted that the information gained through the recommended approach will provide valuable guidelines for pursuing the long range objective of developing a phenomenological understanding of the material removal process.

The important input conditions, controlling factors, operating factors and output response of the material removal process are shown in Figure 2. The input consists of the machining conditions such as speed, feed, depth of cut, etc. The controlling factors which dictate the constraints as well as the operating region are those resulting from the machine tool, cutting tool, workpiece material and cutting fluid used. These controlling factors can be characterized quantitatively in terms of several pertinent aspects. The process is also guided by the operating factors such as an economic objective in terms of cost and production rate. The output of the process is in terms of machining response such as tool life, material removal rate (cutting rate), cutting forces surface roughness, dimensional tolerance and surface integrity.

The key factors in this approach are:

1. Mathematical description of the experimental space in terms of the range of variables such as speed, feed, workpiece hardness, etc.

2. Planning of the least number of statistically valid experiments within the experimental space.

3. Development of a methodology for deterministic, statistical and/or probabilistic mathematical models relevant to material removal processes.

## A Review of Empirical Models

Since the introduction of first emperical tool life model by F.W. Taylor (1900's), the following different categories of models have been introduced.

## Deterministic Models

Historically, the deterministic models appeared from about 1900 to the later 1960's. In this era, the following models were proposed: (a) extended-Taylor; (b) second order after logarithmic transformation, and (c) nonlinear models. During this period, model fitting was primarily conducted graphically.

## Statistical Models

Although the tendency of tool life data to exhibit considerable scatter was recognized prior to the mid-1950's, researchers in the period 1950-1970 began to apply statistical methods to measure the degree of variability. Also, the techniques of least-square fits were applied to generate the model from the experimental data. Importantly, the statistical methods such as factorial and fractional-factorial experiments and variance analysis to draw statistical inference from the fitted model were introduced.

## Probabilistic Models

Since the application of statistical models, especially those for drawing statistical inference, involved an implicit assumption about tool life distributions within the experimental space, researchers began to investigate tool life distributions to verify this assumption. Recognition of the fact that a tool's life may end either by a gradual build up of wear or by a catastrophic failure of the cutting edge such as chipping, fracture, etc., brought forth probabilistic models that contained a tool wear function as well as a hazard function.

## Stochastic and Dynamic Models

The stochastic nature of certain material removal phenomena such as grinding abrasive characterization, chatter and vibration, characterization of machined and ground surface topography, etc., was recognized during the early to mid-1960's. Models of this type were based on either discrete or parametric time series analysis through auto correlation functions or spectral analysis.

Some of the important empirical models are shown in Table II.

## Application of Experimental Designs to Tool Life And Other Material Removal Experiments

Statistical experimental designs have been successfully applied to agricultural, chemical and petrochemical fields since the early 1930's. In the mid-1960's these techniques, including factorial, fractional-factorial and composite designs and response surface methodology, were applied to tool life experiments by Wu and others (1964-1977). Formerly, empirical tool life models (mostly deterministic) were developed through the classical experimental approach in which all variables except one were held at predetermined constant values during the experiments. This approach requires prohibitively large numbers of tests when several variables are to be included in the model. The statistical

TABLE I  Historical Developments of Material Removal Technology

| YEAR | EMPERICAL | PHENOMENOLOGICAL | CUTTING TOOLS | MATERIALS | MACHINE TOOLS |
|---|---|---|---|---|---|
| | | CHIP FORMATION | CARBON TOOL STEELS | | BELT DRIVEN MACHINE |
| 1900 | TAYLOR'S TOOL LIFE MODEL | | HSS TOOL STEELS | BESSEMER STEELS | MOTOR DRIVEN |
| 1940 | EXTENDED TAYLOR MODEL | MECHANICS MODEL | CARBIDE TOOLS | LEADED STEELS | |
| 1950 | ECONOMIC MODELS | PLASTICITY MODEL | | SELENIUM TELLURIUM TREATED STEELS | |
| | | QUICK STOP OBSERVATIONS | CERAMIC TOOLS | TITANIUM | NC MACHINE TOOLS |
| 1960 | STATISTICALLY DESIGNED EXPERIMENTS | TOOL WEAR THEORIES | SUPER HSS TOOLS | NICKEL BASED ALLOYS | MACHINING CENTERS |
| | 2ND ORDER MODELS | SCANNING ELECTRON MICROSCOPIC OBSERVATIONS | COATED CARBIDE TOOLS | CALCIUM DE-OXIDIZED STEELS | DNC MACHINE TOOLS CNC MACHINE TOOLS |
| 1970 | NONLINEAR MODELS | BUE OBSERVATIONS | BORAZON TOOLS | HIP'ED ALLOYS | AC MACHINE TOOLS |
| | OPTIMIZATION MODELS | TOOL WEAR OBSERVATIONS | INDUSTRIAL DIAMOND TOOLS | HIGH SI ALUMINUM ALLOYS | |
| | AC ALGORITHMS | | P/M HSS TOOLS | | CAM SYSTEMS |
| | METHODOLOGY OF MATH MODELS | | | | COMPUTER-AIDED PRODUCTION SCHEDULING |
| 1977 | ECONOMIC MODELS FOR PROCESS PLANNING | | | | COMPUTER-AIDED PROCESS PLANNING |

## TABLE II

### Empirical Models for Tool Life and Other Machining Responses

#### Deterministic

1. Taylor, $\ln T = b_0 + b_1 (\ln V)$

2. Extended-Taylor, $\ln T = b_0 + b_1 (\ln V) + b_2 (\ln F) + b_3 (\ln D)$

3. Second Order, $\ln T = b_0 + b_1 (\ln V) + b_2 (\ln F) + b_3 (\ln D)$
$$+ b_{11} (\ln V)^2 + b_{22} (\ln F)^2 + b_{33} (\ln D)^2$$
$$+ b_{12} (\ln V)(\ln F) + b_{13} (\ln V)(\ln D) + b_{23} (\ln F)(\ln D)$$

4. König-DePiereux, $\ln T + b_0 + b_1 V^{b_2} + b_3 F^{b_4}$

5. Gorki/USSR, $T = T_o \, \mathrm{Exp} \left\{ a \left[ 1 - (1 - b \, V/V_o) \right]^{1/2} \right\}$

#### Statistical

Wu and others, $\underline{Y} = \underline{X}\,\underline{b} + \underline{\varepsilon}$, $\underline{Y}$, $\underline{X}$, $\underline{b}$ and $\underline{\varepsilon}$ are matrices of response, independent, error and estimated parameters.

#### Probabilistic

Brown (1962), Normal distribution, $f(t) = (1/\sigma\sqrt{2\pi}) \, \mathrm{Exp} -\left[ (t-\mu)^2/2 \right]^2$

Log normal distribution has been suggested by several researchers.

Rossetto and Levi (1976), Wear and catastrophic tool failure distribution.

$$f(t) = W_1 \left[ 1 - \mathrm{Exp}(-\lambda t) \right] + (1 - W_1) \int_0^T (\sigma t \sqrt{2\pi})^{-1} \mathrm{Exp} -\left[ (\ln T - \mu)^2 / 2\sigma^2 \right] dt$$

$W_1$ = Weighting constant.

Ramalingam and Watson (1977)

Catastrophic tool failure (random): $f(t) = \frac{1}{\lambda} \mathrm{Exp}(-t/\lambda)$; $= \lambda$ mean

Catastrophic time dependent: $f(t) = \beta t^{\beta-1} / \lambda^{\beta}$ = parameter

# MODELS FOR MATERIAL REMOVAL PROCESSES

methods for planning tool life experiments provide a reliable and interpretable model with a minimum number of experiments.

Problems That Needed to Be Solved

Based on a review of the literature and on experience in applying empirical models and statistically designed experiments, the following problems were identified at the outset of this investigation:

1. Most models have been developed on the basis of laboratory tests, primarily designed to illustrate the specific model building approach. The model building methodology that can be applied to production tool life has not yet been reported in the literature.

2. Statistical inference techniques applicable to the use of tool life and other machining response models have not yet been fully developed.

3. Although much of the theoretical ground work for mathematical models applicable to material removal processes has been laid, comprehensive experimental data and practical applications of the models are presently lacking.

4. Since machining response such as tool life is not known a priori, often the preselected positions and ranges of machining variables such as feed, speed, and depth, as required by the standard factorial, fractional-factorial and composite designs, have led to tool life that is either too short, (hence impractical), or too long (hence too expensive) to investigate.

5. Unlike agricultural or chemical fields where there is little difficulty in defining the working region, the shape of the working region is defined by the response as well as the constraints imposed by the cutting tool, machine tool, and part design specifications. This region is generally highly distorted within the speed-feed-depth variable space.

## A METHODOLOGY FOR THE DEVELOPMENT OF EMPIRICAL MODELS OF MATERIAL REMOVAL PROCESSES

The methodology presented in this paper was developed through several tool wear, tool life and cutting force experiments and an analysis of data reported elsewhere.[1,2] A detailed account of the experiments and analysis as well as the literature review is given in the AFML report (AFML-TR-77-154). In the following a brief summary of the methodology is presented.

## A Methodology

The methodology that evolved at the conclusion of the investigation consists of the following stages:

### Stage I: Design of Initial Machining Experiments

In Stage I, the primary objective is to obtain a viable set of data to initiate mathematical modeling.

- General location in the variable space
- Number of initial tests to be run
- Range of each variable to initially examine
- Test pattern to be employed

A good starting point for the initial testing region may be conditions recommended by a machinability handbook or experience with a similar application. Experimental designs, such as two-level factorials or fractional factorials are recommended to obtain some indication of the overall feasibility region. The risk of too short or too long tool life tests is limited at this early stage.

### Stage II: Feasibility Region Construction

Stage II involves the establishment of (a) part requirement constraints including integrity and dimensional stability (cutter deflection); (b) constraints on the physical limitations of the process including maximum or minimum speed, feed, or cutting horsepower; and (c) economic constraints including maximum and minimum tool life or a minimum metal removal rate. The important considerations are:

- The region of investigation is generally smaller than that used for unconstrained investigations. Thus, the experimental model building procedure may employ simpler model forms and require fewer tests within the region.

- The constrained region is generally of an irregular shape and may change markedly in shape and orientation as tests are added sequentially.

- The constrained variable space is often difficult to visualize, particularly in three or more dimensions.

## Stage III: Design of Optimal Machining Experiments for Precise Mathematical Modeling

Once the feasibility region has been initially constructed, emphasis may then shift to the objective of precise parameter estimation and variance reduction. Recently, the D-optimal criterion was introduced as an experimental objective function for precise parameter estimation of tool life models.[3] The criterion is equivalent to selecting tests at points of maximum variance of the predicted response. In this way, the maximum variance of the predicted response within the region is continually minimized. Confidence intervals are used to define the constraints of the feasibility region (see Figures 3 and 5).

## Stage IV: Working Region Construction

Stage IV involves three major tasks: (i) construction of the working region, (ii) process optimization within this region, and (iii) data collection/model(s) updating during production. The construction of the working region involves probabilistic definition of the constraints in terms of prediction intervals (see Figures 4 and 6). Process optimization within this region can be conducted by applying either mathematical programming techniques or the R-T-F concept of tradeoff functions. This concept is discussed elsewhere.[4] During production, the working region and the mathematical model should be continually updated based on the most current production performance data.

## ECONOMIC MODELS FOR MATERIAL REMOVAL PROCESSES

Economic models used for financial and accounting purposes are not generally suitable for analyzing individual manufacturing processes because these models stop at the departmental or cost center level. Process planning economic models must be capable of evaluating alternatives on individual work centers. Fortunately, a framework for building such models already exists in classical machining economics where time and cost relationships in terms of individual operating conditions have been derived for a variety of machining operations. The classical approach, however, does not lead to efficient economic models because (1) a separate economic model is needed for each operation; (2) only material removal operations can be treated; (3) optimization is limited to one variable at a time and (4) the tool life relationships are generally assumed to be deterministic.

The models presented in this paper are, therefore, based on a recently introduced generalized economic model which overcomes all of the noted shortcomings of the classical approach.[4] A macro-economic model suitable for cost estimation and preplanning

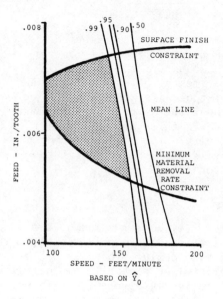

Figure 3. Confidence Interval Regime

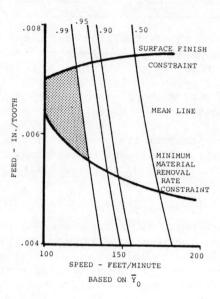

Figure 4. Prediction Interval Regime

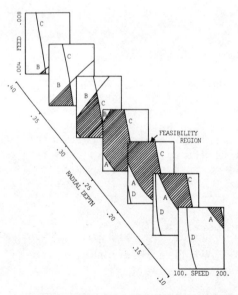

Figure 5. Feasibility Region for End Milling (see data in Ref. 1).

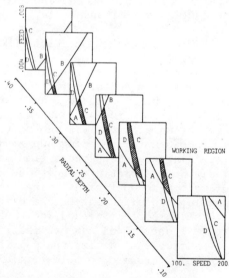

Figure 6. Working Region for End Milling (see data in Ref. 1).

MODELS FOR MATERIAL REMOVAL PROCESSES 269

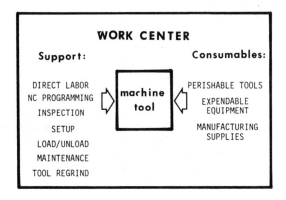

Figure 7. Work Center - An Identifiable Unit for Discrete Parts Manufacturing

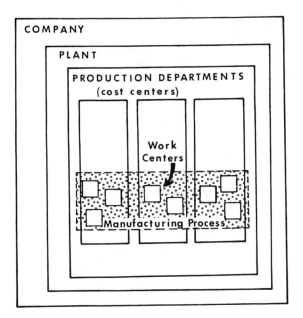

Figure 8. Manufacturing Process for Discrete Parts Processed Through Several Cost Centers and Departments within a Plant of a Company

and a micro-economic model applicable to detailed process planning have been developed and applied to material removal operations on an airframe structure.

Formulation of the Economic Models

Economic models developed for discrete parts manufacturing processes are based on the following observations:

1.  A typical discrete parts manufacturing process such as that used in aerospace companies consists of several individual work centers (e.g., machine tools, heat treating facilities, forging presses, inspection stations, etc.) which are scattered over one or more production departments within a plant or within several plants of a company or several companies. Figure 8 shows a schematic diagram of a manufacturing process.

2.  The productive time within which a work center performs its processing is composed of: (a) setup load/unload time; (b) processing time and (c) time lost in process interruptions for adjustments, tool changes, etc. The economic models described in this paper are concerned with the productive time only. Appropriate equipment and personnel utilization factors must be added to the productive time if estimates of floor-to-floor or lead times are required. The floor-to-floor time is productive time + allowed slack time. To compute lead time in addition to productive time and slack time, nonproductive time (such as that spent in unworked shifts, vacations, holidays, etc. should be included).

3.  The cost elements associated with the productive times of each of the work centers of a manufacturing process are obtained by multiplying each of the components of the productive times by the work center rate, M, which consists of appropriate labor and overhead burden rates, and the appropriate direct and indirect material and auxilliary costs associated with that work center. A typical set of variable cost factors associated with a work center are shown in Figure 7. Computation of the work center rate for each work center in the manufacturing process is an important task. Reconciliation of these rates with the departmental overheads is desirable for a direct tie-in with accounting.

The generalized economic function,

$$I = \lambda_0 + \lambda_1/R + \lambda_2/RT + \lambda_3 \tag{1}$$

is assumed valid for each work center. Time and cost models for each work center are:

$$t = m_0 + m_1/R + m_2/RT + m_3 \qquad (2)$$

$$c = k_0 + k_1/R + k_2/RT + k_3 \qquad (3)$$

The parameters and coefficients of equations (1), (2) and (3) are defined in the nomenclature.

## Macro-Economic Models

Process alternatives for the manufacture of an aerospace part are evaluated during the pre-planning activity of process planning. For any given part, there may be numerous satisfactory alternatives including different machine tools, cutting tools, tool sequences, cutter paths, and feeds and speeds. Regardless of the operation selected, however, there exists one set of alternatives that will result in the most cost effective manufacturing sequence. The macro-economic model enables a pre-planner to evaluate the cost and time for each alternative and to select the most cost effective manufacturing processes.

The time and cost equations (2) and (3) are reorganized for obtaining the following macro-economic model form:

$$t = m_0 + m_1/H_t \qquad (4)$$

$$c = k_0 + k_2/H_c \qquad (5)$$

or $\quad c = (t)(M) + k_3 \qquad (6)$

where t and c are time and cost per part for each operation and $H_t$ and $H_c$ are productivity functions.

$$H_t = R/(1 + m_2/m_1 T) \qquad (7)$$

$$H_c = R/(1 + k_2/k_1 T) \qquad (8)$$

The metal removal rate, R, is characteristic for a certain operation on a particular machine tool with a particular cutter and workpiece material. At this metal removal rate, the tool will last for T minutes. Values for R, T, $H_t$ and $H_c$ are collected for all operations and all alternatives for a particular part family based on the most uncomplicated part geometry. This data comprises the Base Machinability Data.

$H_t$ and $H_c$ must be adjusted for a particular part geometry to reflect the complexity of the part. This is accomplished through the use of cost drivers and cost driver functions. Cost drivers are characteristics of a part which slow down the productivity of the process. They include complexities such as surface finish, tolerance, part rigidity, material variations, etc. The cost driver function is the mathematical relationship which describes the reduction in $H_t$ and $H_c$ for a given cost driver.

## Micro-Economic Model

Economic evaluation at the micro level assumes that detailed information is available about all operations to be considered. For example, detailed cutter paths, tool life information and part configuration must be known. In addition, detailed data must be available for all machine tools, cutting tools, material specifications and cutting fluids.

The micro-economic model is based on the generalized time and cost equations (2) and (3) for each work center. To apply these equations to material removal operations, they are rewritten in the form of the operating variables below:

$$T = m_0 + m_1/R + m_2/RT + m_3$$

$$= (t_s/n + t_L + l_r/r) + (v + v_a)/R$$

$$+ vt_d/RT + m_3 \tag{9}$$

$$c = k_0 + k_1/R + k_2/RT + k_3$$

$$= (t_s/n + t_L + 1/r)M + (v + v_a)M/R$$

$$+ (Mvt_d + Y)/RT + k_3 \tag{10}$$

where $R = R(V,F,AD,RD)$ and $T = T(V,F,AD,RD)$

## Interface Between Macro- and Micro-Economic Models and Other Stages of Discrete Parts Manufacturing

The macro- and micro-economic models can be used independently. However, their full impact is realized when they are interfaced with each other and with the other stages of manufacturing as shown in Figure 9. In the interfaced mode, the data and decisions follow in the direction of the arrows shown. This concept works because the generalized economic model is common to the macro- as well as the micro-stages, so that the data base can be shared between them. The data at the coarse level is related to and based on the specific data either from production feed back or from detailed process planning. For example, at the macro-stage, cutting rate and tool life may be only nominal values. At

# MODELS FOR MATERIAL REMOVAL PROCESSES

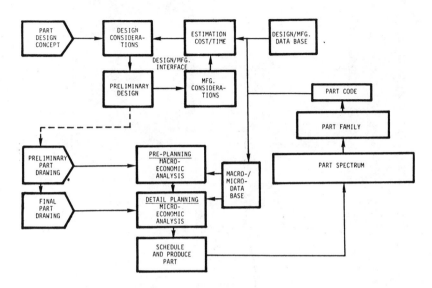

Figure 9. Overall Process Planning System Block Diagram

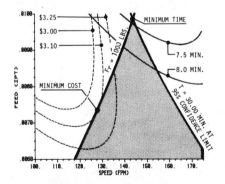

Figure 10. Definition of Working Region for End Milling (see data in Ref. 2).

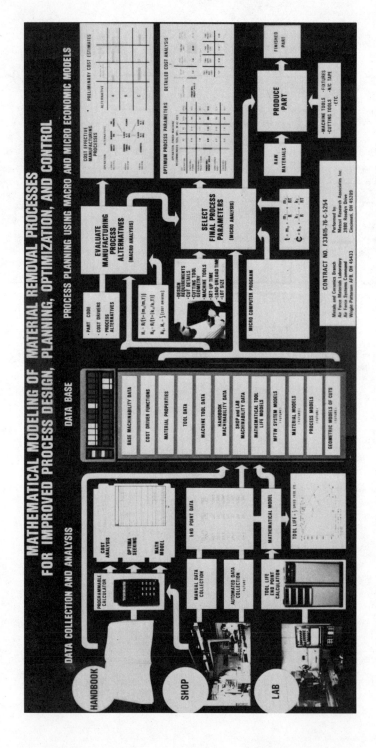

Figure 11. Interrelationships between Data Collection and Analysis, Data Base and Economic Models

the micro- stage, however, their values will be within the working region defined by the cutting rate, tool life and constraint functions.

## Application of Micro-Economic Model

Application of micro-economic model to detailed process planning of an airframe part is discussed elsewhere.[2] To demonstrate the potential of this model, the contour plot of constant cost and production time lines with the resultant cutting force constraint of 1000 lbs. and the 95% confidence limit for 30 minute tool life is shown in Figure 10. The working region for acceptable parts is shaded. By examining the cost and production time contours, the minimum time and cost within the qualified area is established. The cost contour which is tangent to the 1000 lb. force contour is the $3.10 contour. This represents the minimum cost in the working region. The minimum production time contour of 7.50 minutes in the working region passes through the intersection of the 1000 lb. force contour and the 30 minute (95% confidence) tool life contour.

A comparison of the optimized cost with the estimated cost for the cut shows that almost 300% reduction is possible ($3.10 versus $9.79).[2] This clearly illustrates how optimization of the process using mathematical models for tool life and cutting force provides a significant improvement over the use of starting recommendations.

## CLOSING REMARKS

The schematic diagram shown in Figure 11 describes the overall interrelationships among the various phases of the "data-based" system: data collection and mathematical model, data base, macro-economic model and cost estimation, and micro-economic model and parameter optimization. An important feature of the system is the flow of data through analysis and modeling stages before it enters the economic models for evaluation of process alternatives and parameter optimization.

## ACKNOWLEDGEMENTS

The concepts presented in this paper were developed under Air Force Conctract No. F33615-76-C-5254. It is a pleasure to acknowledge contributions from Messrs. Steve Buescher, Steve Vogel, John Christopher, Ray Garrison of Metcut Research and Professor R.E. DeVor and Mr. William Zdeblick of the University of Illinois. Ms. Susan Harvey, Betty Bryk and Barbara Cardelli deserve a special note of thanks for the editing and typing of the report and technical manuscript.

REFERENCES

1. "Development of Mathematical Models for Process Planning of Machining Operations," R.E. DeVor, W.J. Zdeblick, V.A. Tipnis, and S. Buescher, Proc. of Sixth NAMRC, April 1978.

2. "Economic Models for Process Planning," V.A. Tipnis, S. Vogel, and H.L. Gegel, Proc. of Sixth NAMRC, April 1978.

3. "An Experimental Strategy for Designing Tool Life Experiments," W.J. Zdeblick and R.E. DeVor, Journal of Engrg. for Industry, Trans. ASME, Series 3, 77-WA, Prod-24, 1977.

4. "Cutting Rate-Tool Life Function (R-T-F): General Theory and Applications," G.L. Ravignani, V.A. Tipnis, and M.Y. Friedman, CIRP General Assembly, August 1977, New Delhi, India.

*Currently President of Tipnis Associates, Inc., Cincinnati, Ohio.

MODELLING MACROSEGREGATION IN

ELECTROSLAG REMELTED INGOTS

D. R. Poirier*, M. C. Flemings**, R. Mehrabian***, and H. J. Klein****

*University of Arizona, Tucson, Arizona
** Massachusetts Institute of Technology, Cambridge, Ma
*** University of Illinois, Urbana, Illinois
**** Stellite Division, Cabot Corp., Kokomo, Indiana

ABSTRACT

Equations for predicting flow of interdendritic liquid and macrosegregation in ingots prepared by electroslag remelting (ESR) are derived, and computer predictions of macrosegregation based on these equations are compared with macrosegregation measured in experimental ingots.

Agreement between calculations and experimental results is excellent. Experiments have been on model ESR ingots (65-105 mm diameter) of Sn-Pb and Al-Cu alloys; in addition, results are discussed for a nickel-base alloy produced in a laboratory ESR ingot mold with a diameter of 200 mm. The influence of the important solidification parameters such as the shape and depth of the mushy zone and the local solidification time on the macrosegregation across the ESR ingots is quantitatively demonstrated. It is shown that macrosegregation theory predicts not only surface-to-center variations in compositions, but also predicts conditions under which a severe type of segregation, called "freckles", forms in ESR ingots.

A method of minimizing macrosegregation is demonstrated whereby ingot rotation alters interdendritic flow behavior and therefore macrosegregation. Modest rotational speeds eliminate "freckles" and minimize surface-to-center type segregation, as well. It is also suggested that macrosegregation theory should be considered during the alloy design stage in that alloy constitution can possibly be adjusted to produce ESR ingots with no "freckles" and minimum segregation.

Recommendations to improve the "state of the art" in modelling macrosegregation are given. For example, the effect of convection of the liquid pool should be examined and possibly included in future models. In addition, the effect of the interdendritic liquid flow should be included in the energy equation when it is applied to the mushy zone of large ingots, and possibly, the electromagnetic force field should be included in the equation of motion in the mushy zone. The selections of values of permeability used for macrosegregation simulation are also discussed and compared to permeability measured by experiment.

## INTRODUCTION

This paper summarizes research carried out at the Massachusetts Institute of Technology under sponsorship of the Army Materials and Mechanics Research Center and the first phase of a three-year program between the University of Illinois and Cabot Corporation sponsored by the National Science Foundation. Both programs have been directed towards the electroslag remelting process with particular emphasis put on modelling the flow of interdendritic liquid within the mushy zone and the resulting macrosegregation due to that flow. The aim is to develop a model of the ESR process which would predict the correct remelting parameters for production of large (5 to 50-ton) ingots free from segregation. At the present time, a number of full-scale ESR production runs are normally required to determine the correct processing parameters for each new alloy. Furthermore, while segregation is reduced, it generally persists and can reach unacceptable levels with increasing ingot size.

Although there are a number of important factors which determine the acceptability of an ESR ingot, in general, the most important are ingot structure and homogeneity. The necessity of obtaining the specific desired structure and homogeneity generally is the limiting factor in determining among other things maximum ingot cross-sectional area and ingot melt rate.

Homogeneity is especially important for alloys which are segregation prone such as some nickel-base superalloys, and a number of tool and high alloy steels. Even with careful control of the remelt parameters, segregation sometimes occurs. A model of the process quantifies the parameters which must be controlled to eliminate or minimize segregation during remelting. Such parameters as melt rate, section size, slag depth, fill ratio, etc., are probably more important than now realized. It is expected that minimizing segregation will become even more important as the ingot size increases from the present 500 mm to 1000 mm in diameter to diameters of 2500 mm and greater. For example, the production of ESR ingots for forged round billets of stainless steels and tool steels was found profitable only when ingot diameter is greater than 500 mm and the greater the diameter, the more profitable (1).

The important solidification parameters which determine the severity of macrosegregation are qualitatively understood by producers. Specifically, in addition to alloy composition and ingot diameter, the operating conditions which lead to a deep liquid pool shape cause increased macrosegregation (2,3). Consequently, a number of studies (4,5,6) have been on determining the parameters which control depth of the liquid metal pool. The most important parameter is the melting rate of the electrode with a higher melting rate leading to a deeper metal pool and deeper mushy zone. Other parameters are composition and amount of slag, current and voltage, and electrode polarity.

Heat flow models have been developed to relate shape of liquid pool and mushy zone to the parameters mentioned above (6-10), but none of these models has incorporated the constitutive equations for the flow of interdendritic liquid necessary to predict macrosegregation. The purpose of this paper is to summarize the relevant equations for flow of interdendritic liquid and to show how these equations can be used to calculate macrosegregation (11-13).

## ANALYSIS OF MACROSEGREGATION

In the last decade, work on macrosegregation has shown that flow of solute-rich interdendritic liquid and in some cases mass flow of liquid plus solid in the mushy zone of a casting is responsible for almost all types of segregation observed (11-22). Movement of interdendritic liquid occurs as a result of solidification shrinkage, the force of gravity acting on a liquid of variable density, and penetration of bulk liquid in front of the liquidus isotherm, due to fluid motion in this region, into the mushy zone.

The density of the interdendritic liquid within the mushy zone of a solidifying ingot varies due to constitutional and thermal effects, and because of gravity there is natural convection. We refer to this as "gravity-induced" convection. Convection of interdendritic liquid also occurs due to solidification shrinkage since liquid "feed" metal must flow towards regions where the solidifying solid has a density greater than the local interdendritic liquid. This contribution is called the "solidification induced" or "shrinkage induced" convection.

In the results presented here, the effect of bulk liquid in the liquid metal pool, above the liquidus isotherm, on the interdendritic flow velocity is not considered. The flow in the bulk metal pool is due to the electromagnetic force field and natural convection. Computed results show that the electromagnetic force field is the more important driving force for fluid motion, although the natural convection does affect the circulation pattern (23). The convection is greatest in the slag phase, with a strong circulatory pattern just below the electrode in the region near the mold wall. In the liquid pool, calculations indicate that the convection is weaker than in the slag and is about 30 - 50 mm/s (23). Treatment of this aspect of the flow will be included when the appropriate models are developed. At this time, this effect is thought to be small since the penetration bulk liquid encounters a dense dendritic network, which greatly impedes the bulk flow, at a short distance behind the advancing dendrite tips (i.e., the liquidus isotherm). It is desirable, however, to be able to verify this assumption; such work is planned for the near future.

Historically, "solidification induced" convection was probably first treated by Scheil (24) and Kirkaldy and Youdelis (25), which led to our understanding of the type of segregation called "inverse segregation." However, these analyses are only applicable to unidirectional heat and fluid flow. Somewhat later, Flemings and Nereo (14) showed that macrosegregation could be quantitatively treated for cases other than simple unidirectional and heat and fluid flow and presented the "local solute redistribution equation" (LSRE) which is a key relationship in macrosegregation theory. Because of the importance of the LSRE, we review its development here.

We consider a small volume element within the mushy zone. It is large enough that the volume fraction of solid within it is equal to the local average, but small enough that it can be treated as a differential element as in problems of flow through porous solids. Solute enters or leaves the element only by liquid flow. Mass flow in or out of the element by diffusion is neglected. Liquid composition within the element is uniform and is related to temperature by the liquidus of the appropriate phase diagram. This assumes, of course, that during solidification there is negligible undercooling of the interdendritic liquid due to constitutional and/or kinetic effects which are valid assumptions for dendritic growth of alloys. Also for most alloy elements, we can assume negligible solid diffusion during solidification. With these assumptions, solute redistribution within the element is given by (14).

$$\frac{\partial g_L}{\partial C_L} = - \left(\frac{1 - \beta}{1 - k}\right) \left[1 + \frac{\vec{v} \cdot \nabla T}{\varepsilon}\right] \frac{g_L}{C_L} \qquad (1)$$

where  $g_L$ = volume fraction liquid,

$C_L$ = composition of liquid,

$\beta = (\rho_S - \rho_L)/\rho_S$ (i.e., solidification shrinkage),

$\rho_S$ = density of solid,

$\rho_L$ = density of liquid,

k = partition ratio,

$\vec{v}$ = velocity vector of interdendritic liquid,

$\nabla T$ = local temperature gradient, and

$\varepsilon$ = local cooling rate (rate of temperature change).

In addition to assumptions given above, other assumptions are constant solid density, no pore formation, and no solid moves in or out of the volume element during solidification.

The calculation of the flow of interdendritic liquid due to solidification shrinkage combined with "gravity induced" flow was first performed by Mehrabian et al. (18). In that work, the effect of gravity as a body force on the convecting liquid within the mushy zone was introduced by using D'Arcy's Law, which is

$$\vec{v} = - \frac{K}{\mu g_L} (\nabla P + \rho_L \vec{g}) \qquad (2)$$

where  K = specific permeability,

$\mu$ = viscosity of the interdendritic liquid,

P = pressure, and

$\vec{g}$ = acceleration due to gravity

When flow takes place through the volume element, conservation of mass requires that

$$\frac{\partial}{\partial t}(\rho_S g_S + \rho_L g_L) = - \nabla \cdot \rho_L g_L \vec{v} \qquad (3)$$

where t is time and $g_S$ is volume fraction solid. Equation (3) is the continuity equation written for two phases, solid and liquid, with the assumption that the solid phase is stationary.

In the "local solute redistribution equation," it is assumed that equilibrium exists at the interface between the interdendritic liquid and the solid phase. Therefore, the temperature of a solidifying alloy is dictated by the liquidus line on the corresponding phase diagram, and the composition of interdendritic liquid is a function of temperature only, so

$$\frac{\partial C_L}{\partial t} = \frac{dC_L}{dT} \frac{\partial T}{\partial t} = \frac{\varepsilon}{m} \qquad (4)$$

and

$$\frac{\partial \rho_L}{\partial t} = \frac{d\rho_L}{dC_L} \frac{\partial C_L}{\partial t} = \frac{d\rho_L}{dC_L} \frac{\varepsilon}{m} \qquad (5)$$

where m is the slope of the liquidus on the phase diagram if the alloy is a simple binary. In general, m is $dT/dC_L$, the change in liquidus temperature for a change in $C_L$.

Finally, by combining Equations (1) - (5), we get

$$\nabla \cdot \left(\frac{K\rho_L}{\mu} \nabla P + \frac{K\rho_L^2}{\mu} \vec{g}\right) =$$

$$- (\rho_L - \rho_S) \left(\frac{1-\beta}{1-k}\right) \left[1 - \frac{K}{\mu \varepsilon g_L} (\nabla P + \rho_L \vec{g}) \cdot \nabla T\right] \frac{g_L}{C_L} \frac{\varepsilon}{m}$$

$$+ g_L \frac{d\rho_L}{dC_L} \frac{\varepsilon}{m} \cdot \qquad (6)$$

Equation (6) can be solved along with appropriate boundary conditions to give pressure within the mushy zone of a solidifying ingot. In addition to specifying appropriate boundary conditions, the temperature field within the mushy zone must be known so that values of $\nabla T$ and $\varepsilon$ can be specified. Alternatively,

# MODELLING MACROSEGREGATION

the temperature field can be determined by solving an appropriate energy equation.

When the pressure is known throughout the mushy zone, Equation (2) is applied to determine local velocity, $\vec{v}$, of the interdendritic liquid. With the obtained velocity, Equation (1) is integrated to establish the distribution of $g_L$ within the mushy zone.

Finally, macrosegregation in an ingot is given in terms of the local average composition of solid after solidification is complete. From Flemings and Nereo (14), this is

$$\bar{C}_S = \frac{\rho_S k \int_0^{1-g_E} C_L dg_S + \rho_{SE} g_E C_E}{\rho_S (1 - g_E) + \rho_{SE} g_E} \tag{7}$$

In Equation (7), $g_E$ is the volume fraction of eutectic and $C_E$ is the eutectic composition. The integration in the numerator can be carried out by picturing a fixed volume element in which $C_L$ continuously changes in a known manner as the alloy solidifies. For an ESR ingot, when the isotherms move at a steady velocity, the integration can be carried out by integrating from the liquidus ($g_S = 0$) down to the solidus ($g_S = 1-g_E$) at a given radius. By doing this for different radial positions within the ingot, $\bar{C}_S$ versus radius is determined which can be plotted to give the pattern of macrosegregation.

To sum up, the analytical model (given a temperature distribution) can be used to predict:

(1) the pressure distribution within the mushy zone, Equation (6);

(2) the velocity of interdendritic liquid flow within the mushy zone, Equation (2);

(3) the distribution of volume fraction liquid within the mushy zone, Equation (1); and

(4) the local average composition after solidification is complete, Equation (7).

## CYLINDRICAL REMELTED INGOTS

The above analysis was first given and applied by Mehrabian et al. (18). They calculated pressure, interdendritic liquid flow and macrosegregation in an ingot solidified horizontally with unidirectional heat flow. More recently, the same equations have been applied to cylindrical remelted ingots (11,12) such as that depicted in Figure 1. Specifically, for such an application, Equation (6)

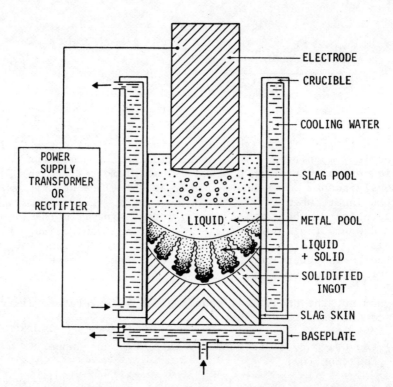

Figure 1: Schematic illustration of an axi-symmetric ESR system.

is expanded into cylindrical coordinates to arrive at the pressure in the mushy zone of a solidifying ESR ingot (11,12):

$$\frac{\partial^2 P}{\partial r^2} + \frac{\partial^2 P}{\partial z^2} + A \frac{\partial P}{\partial r} + B \frac{\partial P}{\partial z} + C = 0 \quad (8)$$

where A, B and C are defined as follows:

$$A = \frac{1}{r} + \frac{2}{g_L}\left(\frac{\partial g_L}{\partial r}\right) + \frac{1}{\rho_L}\left(\frac{\partial \rho_L}{\partial r}\right) + \alpha\left(\frac{\partial C_L}{\partial r}\right), \quad (9)$$

$$B = \frac{2}{g_L}\left(\frac{\partial g_L}{\partial z}\right) + \frac{1}{\rho_L}\left(\frac{\partial \rho_L}{\partial z}\right) + \alpha\left(\frac{\partial C_L}{\partial z}\right), \quad (10)$$

$$C = g\rho_L \left[\frac{2}{g_L}\left(\frac{\partial g_L}{\partial z}\right) + \frac{2}{\rho_L}\left(\frac{\partial \rho_L}{\partial z}\right) + \alpha\left(\frac{\partial C_L}{\partial z}\right)\right] - \frac{\varepsilon\mu}{m\gamma g_L}\left[\frac{1}{\rho_L}\frac{d\rho_L}{dC_L} + \alpha\right] \quad (11)$$

with

$$\alpha = \frac{\beta}{(1-k)C_L} \ .$$

The boundary conditions are shown in Figure 2. Since the mold wall is impermeable, $v_r = 0$ at the wall, and at the centerline, $v_r = 0$ because of symmetry. At the solidus isotherm, the boundary condition satisfies continuity of the solidifying eutectic liquid (18); $v_r$ and $v_z$ are the velocity components of the eutectic liquid, $U_{Er}$ and $U_{Ez}$ are the velocity components of the solidus isotherm (eutectic temperature), and $\rho_{SE}$ and $\rho_{LE}$ are the densities of eutectic solid and liquid, respectively. Within the bulk liquid pool, we assume no convection; $P_0$ is the pressure at the top of the liquid pool, $\rho_{LO}$ is the density of the bulk liquid, and h is the height of the liquid pool.

From the measured shape of mushy zone, temperature distribution in the mushy zone, solidification rate and cooling rate, all the unknown variables (except $g_L$) involved in A, B and C of Equation (8) can be determined with the help of a phase diagram and a density versus liquid composition diagram. To initiate calculations $g_L$ is approximated using the Scheil Equation (i.e., Equation (1) with $\beta$ and $\bar{v}$ equal to zero). Now, with the boundary conditions given in Figure 2, Equation (8) can be solved for the pressure

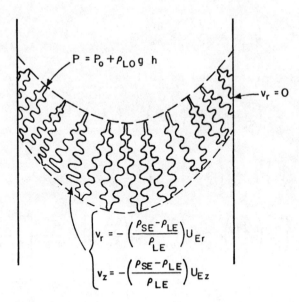

Figure 2: Boundary conditions used in solving for flow of interdendritic liquid.

distribution in the mushy zone. Once the pressure distribution is known, the velocity of interdendritic liquid in the mushy zone is calculated using D'Arcy's Law, Equation (2). With the obtained velocity distribution, we integrate Equation (1) to obtain new values of $g_L$ which are substituted into A, B and C to recalculate a new pressure and velocity distribution. This procedure is repeated until $g_L$ stops changing. This final $g_L$ distribution is the correct one. Finally, with this correct $g_L$ distribution in the mushy zone, the local average composition, $C_S$, can be calculated by using Equation (7).

When flow of interdendritic liquid is relatively extensive, resulting in pronounced segregation, as in the ingot of Figure 3, it is important to numerically compute $g_L$ since the solute accumulation due to the flow of interdendritic liquid can be significant. For example, Figure 4 shows that the discrepancy in solute

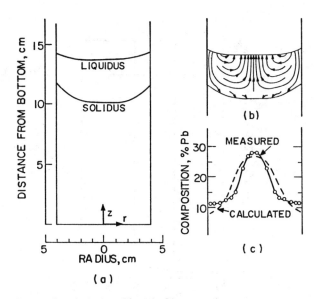

Figure 3: Interdendritic fluid flow and macrosegregation in a simulated ESR ingot (Sn-15% Pb). (a) Steady-state liquidus and solidus isotherms; (b) flow lines of interdendritic liquid; (c) macrosegregation measured and calculated. From Reference 12.

accounting between the case of no interdendritic flow (Scheil equation) and with flow is significant along the center of the ingot of Figure 3.

## EFFECTS OF SOLIDIFICATION PARAMETERS

Macrosegregation has been calculated, using the above relationships, for several small-scale experimental ingots for Sn-15% Pb and Al-4% Cu alloy and compared with experimental results (11, 12). Typical results are shown in Figure 3(c), which shows that calculations and experiment compare remarkably well. In these

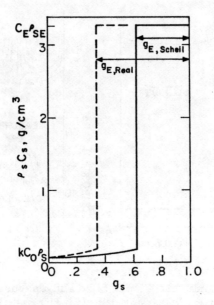

Figure 4: Solute accumulation with interdendritic liquid flow and neglecting flow (Scheil equation). From Reference 12.

studies, solidification was monitored with thermocouples so that the temperature field was known (e.g., see Figure 3(a) and served as input to macrosegregation calculations. Details of the apparati employed can be found in the original references (11,12,26). These experiments have fortified the applicability of quantitatively calculating flow of interdendritic liquid in the mushy zone of a solidifying remelted ingot and using that calculation as a basis for predicting macrosegregation.

The effect of vertical solidification rate and depth of mushy zone on macrosegregation can then be studied as illustrated in Figure 5. Here, macrosegregation has been calculated for small

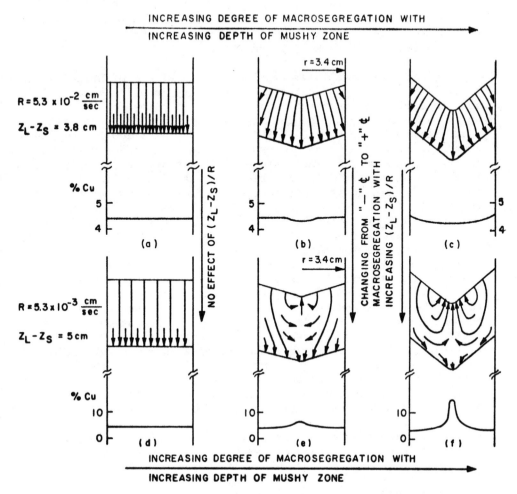

Figure 5: Effects of mushy zone shape and vertical solidification rate on macrosegregation in Al-4.4% Cu. From Reference 12.

Al-4.4% Cu ESR ingots for different solidification conditions. Calculated flow lines and macrosegregation are shown for two different vertical solidification rates and for mushy zones of three different degrees of concavity (i.e., three different "depths" where depth refers to distance from the highest to the lowest point in the mushy zone). Note that macrosegregation increases with increasing depth of mushy zone. At the higher solidificaion rate (smaller local solidification time), it is negative at the centerline and at the slower solidification rate, it is positive at the centerline. At these slower solidification rates, gravity causes the interdendritic flow lines to curve inwards and upwards near the center (Figures 5(e) and (f)).

In addition to illustrating the strong effect of solidification rate, Figure 5 also indicates the difficulty in using small ESR ingots to study macrosegregation in commercial alloys before scaling up to full-size production ESR ingots. The usual experience is that alloys made into small ESR ingots, typically 200-250 mm diameter, show little or no macrosegregation. The ingot depicted in Figure 5(c) was, in fact, produced in a small-scale ESR unit. Flow of interdendritic liquid primarily satisfies solidification contraction, and so it must diverge and flow from hotter to colder regions in the mushy zone results in slight negative segregation at the center. Under such conditions, the severe localized segregates (often called "freckles") do not form. On the other hand, when the vertical solidification rate, R, is decreased by an order of magnitude (hence, local solidification time is increased by an order of magnitude), flow is strongly driven by the gravity force field. The more dense and solute-rich liquid flows downward and toward the center and then upward from cooler to hotter regions in the mushy zone resulting in strong positive segregation, Figure 5(f). When the flow reversal is sufficiently strong, "freckles" will form. The key to experimental studies on macrosegregation in ESR ingots is to use equipment which will simulate the vertical solidification rate of production ingots (approximately $10^{-2}$ mm/s). Designs, at least suitable for alloys with low melting temperatures, are shown in Ridder et al. (11) and Kou et al. (12).

It is also possible to experimentally verify macrosegregation theory for high temperature alloys produced in ESR ingot molds not equipped with thermocouples. In this approach, metal pool profiles are marked by successive additions of tungsten powder. An example of this is shown in Figure 6. Successively marked metal pools permit determination of the vertical pool velocity, R. Ingots thus produced can be sectioned and their dendritic arm spacings versus distance from the mold wall determined. Mushy zone thickness and shape during ESR are then deduced utilizing the data for secondary dendrite arm spacing versus local solidification time or average cooling rate (11,12,27,28,29). For example, it can be shown that

$$z_L - z_S = \Delta T \cdot R \cdot (d/a)^{1/n} \qquad (12)$$

where $z_L - z_S$ = height of the mushy zone at a given radial position in the ingot, r;

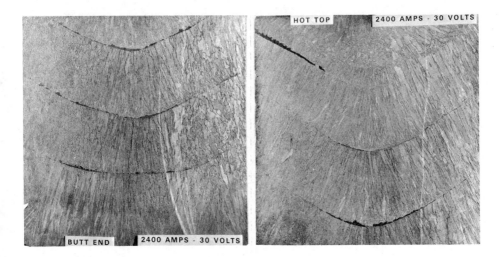

Figure 6: Metal pool profiles in an ESR ingot of Ni-27% Mo alloy marked by tungsten powders. Ingot and electrode diameters, ingot height, slag thickness, melt rate and average vertical pool velocity were 196mm, 152mm, 400mm, 57mm, 26.2 g/sec, and 0.09 mm/sec, respectively. From Reference 11.

$\Delta T$ = liquidus minus the nonequilibrium solidus temperature of the alloy;

d = measured secondary dendrite arm spacings at r; and

a,n = constants.

The constants, a and n, are predetermined for a given alloy using data of secondary dendrite arm spacing, d, versus average cooling rate as follows:

$$d = a \left[ \frac{\partial T}{\partial t}_{avg} \right]^{-n} \tag{13}$$

Figure 7 shows the liquidus and solidus isotherm at two locations during production of an Ni-27% Mo alloy ingot determined from tungsten markings, dendrite arm spacing measurements, and Equation (12). The calculated interdendritic flow velocities are shown in

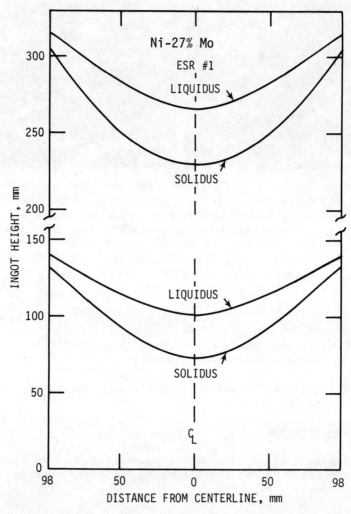

Figure 7: Solidus and liquidus isotherms at two locations during production of a Ni-27% Mo alloy. ESR ingot Tungsten markings are shown in Figure 16. Solidus isotherm was determined from dendritic arm spacing measurements and Equation (12). From Reference 11.

Figure 8. The value of R for this ingot was ~ $10^{-1}$ mm/sec. Due to this, along with a rather narrow mushy zone size and the small liquid density gradients, there was not significant interdendritic flow due to the density differences. The value of $\gamma = 1.5 \times 10^{-5}$ mm² was used in the calculations. The direction of flow in Figure 8 diverges slightly outward in this case and small negative segregation at the ingot centerline is expected. The experimental and theoretical segregation curves for this 200 mm ingot are shown in Figure 9.

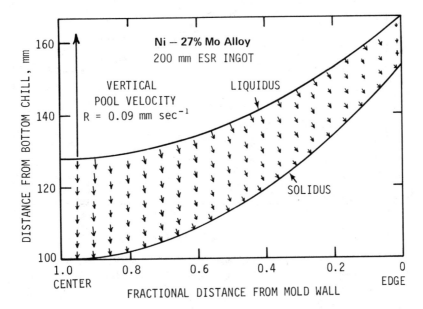

Figure 8: Calculated flow velocities in a 200 mm diameter Ni-27% Mo ESR Ingot. The arrows designate the direction of interdendritic fluid flow. The length of each arrow denotes its magnitude. Note that flow is consistently from the hotter to the cooler region. From Reference 11.

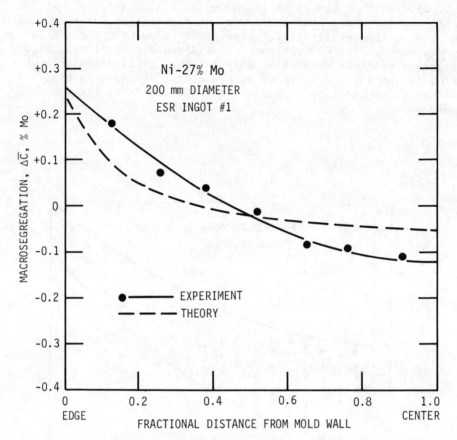

Figure 9: Comparison of experimental and theoretical segregation profiles in the ESR ingot of Ni-27% Mo alloy. Cross sectional area analyzed was 100mm from the bottom of the ingot. From Reference 11.

## FORMATION OF "FRECKLES"

The computer models employed above predict the conditions which cause the formation of a "freckle."

According to macrosegregation theory (18), the important dimensionless parameter affecting macrosegregation is $\vec{v} \cdot \nabla T/\varepsilon$

which appears in Equation (1). When this is equal to $\beta/1 - \beta$ no macrosegregation results; when it is greater, segregation is negative and when it is less, segregation is positive. Alternatively, it can be shown that

$$\frac{\vec{v} \cdot \nabla T}{\varepsilon} = - \frac{\vec{n} \cdot \vec{v}}{\vec{n} \cdot \vec{u}} \qquad (14)$$

where $\vec{u}$ is the isotherm velocity and $\vec{n}$ is a unit vector normal to the isotherm. Then, it is convenient to view the dimensionless parameter, $-(\vec{n} \cdot \vec{v})/(\vec{n} \cdot \vec{u})$ as <u>local flow velocity perpendicular to isotherm relative to isotherm velocity</u>. Macrosegregation criteria can then be established from the following equation (11):

$$1 - \frac{\vec{n} \cdot \vec{v}}{\vec{n} \cdot \vec{u}} \gtreqless \frac{\rho_s g_L + g_E(\rho_{sE} - \rho_s)}{\rho_L g_L} \qquad (15)$$

It is readily seen from Equation (1) that for the case of the equal sign, Equation (1) reduces to the simple differential form of the nonequilibrium segregation equation. Therefore, no macrosegregation results when the equal sign in Equation (15) applies. When the left-hand side of Equation (15) is larger, there is a net flow from the hotter to the cooler region of the mushy zone than that for the case above, and negative segregation results. When the opposite is true, segregation is positive.

An interesting and important effect occurs when

$$1 - \frac{\vec{n} \cdot \vec{v}}{\vec{n} \cdot \vec{u}} < 0 \qquad (16)$$

In this case, fluid flows in the same direction as the isotherms and faster than the isotherms; thus flow is from cooler to hotter regions within the mushy zone. This type of flow results, from Equation (1), not in solidification but in remelting; volume fraction liquid, $g_L$, increases with increasing $C_L$. This is the basic mechanism of formation of channel-type segregates, including "freckles" (11-13,18,19,30,31).

Figure 10 shows "freckles" observed in a commercial nickel-base alloy ESR ingot produced under improper melting conditions. The elemental iron and niobium x-ray maps of a "freckle" are shown

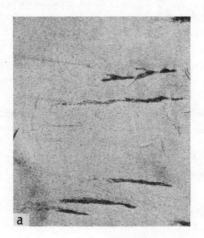
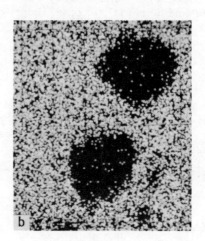
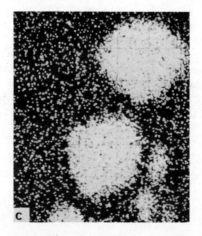

Figure 10: "Freckles" in a commercial ESR ingot of nickel base alloy 718 produced under the improper melting conditions.
    (a) Photograph of the billet at 1.7X
    (b) Elemental Fe X-ray map of the internal structure of a "freckle" at 800X
    (c) Elemental Nb X-ray map of the internal structure of a "freckle" at 800X
(Reduced 10% for Reproduction).

in Figure 10b and 10c, respectively, indicating niobium carbide enrichment of these channels.

Equation (16) can be used to quantitatively predict when and where a channel-type of segregate forms during ingot solidification (32), but after the instability develops (i.e., Equation (16) applies), the calculation of the flow of interdendritic liquid as outlined above is not valid and so a calculation of macrosegregation cannot be made. Figure 11 shows results from a Sn-15% Pb simulated ESR ingot which segregated severely with "freckles" because of the slow vertical solidification rate (7 x $10^{-2}$ mm/s) and very deep concavity of mushy zone. The composition ranges from 8% Pb at mid-radius to 28% Pb at the center of the ingot. Microstructures are shown in Figure 12; "freckles" are evident in Figures 12b and c.

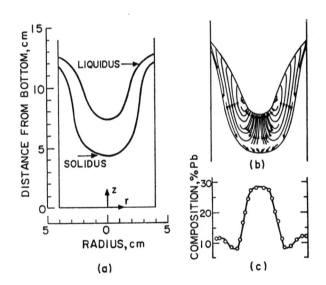

Figure 11: Results obtained for a simulated ESR ingot of Sn-15% Pb alloy. (a) Liquidus and solidus isotherms; (b) calculated flow lines of interdendritic liquid; (c) macrosegregation as measured. From Reference 12.

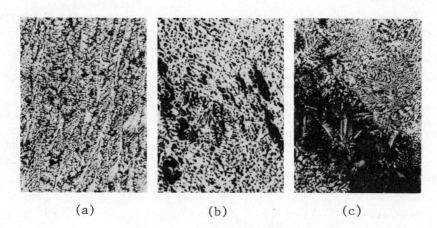

Figure 12: Structure of the ingot of Figure 11: (a) near surface; (b) 15 mm radius; (c) center. Mag. 25.6X. From Reference 12.
(Reduced 10% for Reproduction).

Figure 11 shows calculated flow lines with values of permeability decreased to the point that the flow directions are observed just at the onset of developing a flow instability. Here the value of the permeability is less by almost one order of magnitude than its true value, yet the flow is still toward the centerline and upward at the center and strong enough to satisfy Equation (16) thereby predicting the "freckles" shown in Figure 12.

## SEGREGATION CONTROL

We have seen that when the vertical casting rate, R, is approximately $10^{-2}$ mm/s or less, then macrosegregation is severe and "freckles" can form. This, of course, is due to the gravity force field acting on the interdendritic liquid which varies in density. Here we show the effects of subjecting the liquid to an additional force, centrifugal force, by ingot rotation. The equations which describe the convection of the interdendritic liquid and the macrosegregation across rotated remelted ingots are similar to those employed for stationary electroslag ingots. One difference is that Darcy's Law, Equation (2), now contains a term for the centrifugal force so that

$$\vec{v} = -\frac{K}{\mu g_L} (\nabla P + \rho_L g \vec{z} - \rho_L \omega^2 r \vec{r}) \tag{17}$$

where $\omega$ is the rotational speed.

When Equation (17) is combined with the local solute redistribution equation and the equation of continuity in the mushy zone, we obtain the same equation which describes the pressure distribution inside the mushy zone, namely Equation (8), where A and B are defined as before while C is now defined as follows:

$$C = g\rho_L \left[\frac{2}{g_L}\frac{\partial g_L}{\partial z} + \frac{2}{\rho_L}\frac{\partial \rho_L}{\partial z} + \alpha\left(\frac{\partial C_L}{\partial z}\right) - \alpha\left(\frac{\partial C_L}{\partial r}\right)\frac{\omega^2 r}{g}\right]$$
$$-\frac{\varepsilon\mu}{m\gamma g_L}\left[\frac{1}{\rho_L}\frac{d\rho_L}{dC_L} + \alpha\right] - 2\omega^2\left[\rho_L + r\frac{\partial \rho_L}{\partial r} + \frac{\rho_L r}{g_L}\frac{\partial g_L}{\partial r}\right] \tag{18}$$

Equation (8) is solved for the boundary conditions previously given and shown in Figure 2 except that the pressure at the liquidus isotherm is now given by

$$P(\text{liquidus}) = P_o + \rho_{Lo}gh + \rho_{Lo}\omega^2 r^2/2 \tag{19}$$

The effect of rotating the mold during solidification is shown in Figure 13. In Figure 13a, the ingot when solidified with no rotation ($\omega = 0$), has severe segregation including "freckles" centrally located. When the mold is rotated at a modest rate ($\omega = 8.7$ rad/s, Figure 13b), "freckles" do not result and the macrosegregation is reduced. At a greater speed, Figure 13c, macrosegregation is significantly reduced and the macrosegregation profile approximates a w-shape. However, with excessive rotational speed ($\omega = 15$ rad/s, Figure 13d), the centrifugal force predominates over the gravity force and causes outward flow resulting in freckles at the ingot surface.

The calculations shown in Figure 13 are based upon the data for density given in Figure 14a for the Sn-Pb system. The solid line is for the density of the interdendritic liquid as a function of the composition of Pb up to the eutectic composition where the density of eutectic liquid and solid are shown of $\rho_{LE}$ and $\rho_{SE}$, respectively. The broken line in Figure 14a was used in the

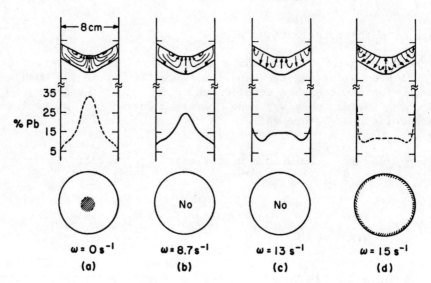

Figure 13: Effect of rotation on flow of interdendritic liquid and macrosegregation in Sn-12.2% Pb ingots. Proceeding from (a) to (d), calculations are for increasing rotational speed. Calculations predict remelting along the center in (a) and at the surface of (d); $R = 3.5 \times 10^{-2}$ mm/s. From Reference 13.

calculations. Calculations were also done for two hypothetical alloys which have the properties of Sn-Pb except that the liquid density is assumed to vary differently during solidification, as shown in Figure 14b. In one case, density of the interdendritic liquid decreases continuously during solidification whereas in the second, density first increases and then decreases. These curves are used for further calculations to show the effect of the manner in which liquid density varies during solidification on convection and macrosegregation in remelted ingots with mushy zones similar to those of Figure 13.

In Figure 15, the alloy of decreasing density is selected as the model. As expected, with no rotation, flow diverges from the center and moves upward toward the mold surface. Indeed, in

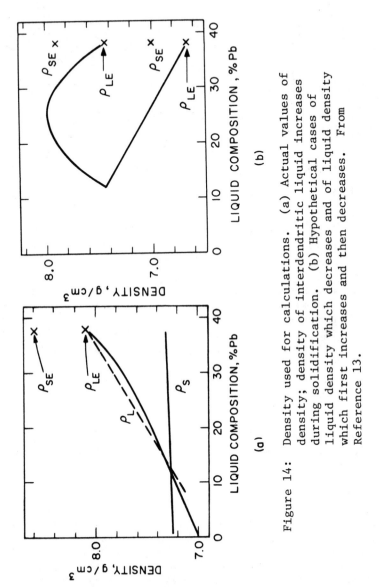

Figure 14: Density used for calculations. (a) Actual values of density; density of interdendritic liquid increases during solidification. (b) Hypothetical cases of liquid density which decreases and of liquid density which first increases and then decreases. From Reference 13.

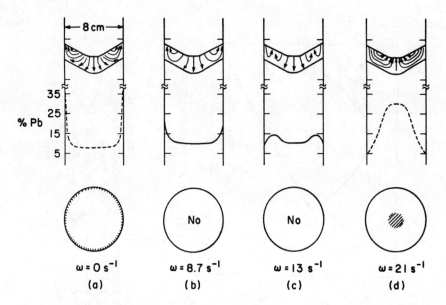

Figure 15: Effect of rotation on flow of interdendritic liquid and macrosegregation in Sn-12.2% Pb ingots assuming a decreasing density of liquid during solidification. Calculations predict remelting at the surface in (a) and along the center in (d); $R = 5.6 \times 10^{-2}$ mm/s. From Reference 13.

Figure 15a, upward flow is so great near the ingot surface that "freckles" form.

Now in this case, the effect of rotating the mold is to shift the positive segregation at the ingot surface to the ingot center, and at a particular speed, the severity of the segregation is minimized. When the rotational speed is excessive (Figure 15d, $\omega = 21$ rad/s), segregation is strongly positive at the center and "freckles" form there in contrast to the situation depicted in Figure 13d. Of the speeds shown, segregation is minimum for $\omega = 13$ rad/s.

The case of liquid density increasing to a maximum and then decreasing is particularly interesting. In nonrotated ingots, Figure 16a, "freckles" appear neither at the center nor at the surface but rather at an intermediate radial position. Since the overall density change is not as great as in the previous two calculations, the degree of segregation is not as severe and the vertical solidification rate must be reduced to $1 \times 10^{-2}$ mm/sec before "freckles" are predicted (Figure 16a). Note that the convection in the lower segment of the mushy zone, where density decreases, is down and toward the center whereas in the upper segment, where density increases, flow is upward.

The effect of rotation in this ingot is to reduce the segregation at the location of the peak, Figures 16b-d, but when the ingot is rotated at an excessive speed (Figure 16d, $\dot\omega$ = 12 rad/s),

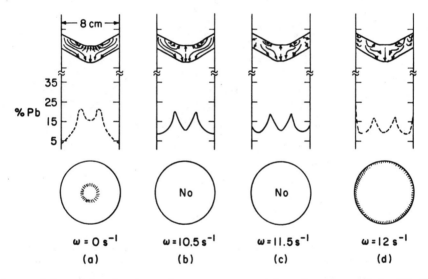

Figure 16:  Calculation of macrosegregation in rotated Sn-12.2% Pb ingots assuming a liquid density which increases and then decreases during solidification. Calculations predict remelting at the peaks in (a) and at the surface in (d); $R = 1.0 \times 10^{-2}$ mm/s. From Reference 13.

the greater centrifugal force at the ingot surface prevails and causes the formation of "freckles" at the ingot surface.

The quantitative effect of the centrifugal force on the macrosegregation is more clearly shown in Figure 17 for the case of the

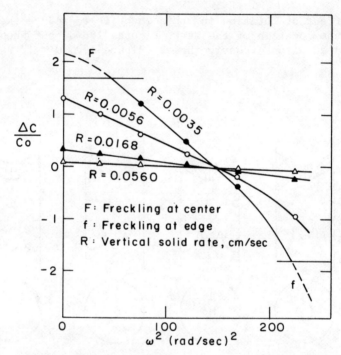

Figure 17: Difference in composition ($\Delta C$) between center and surface as a function of rotational speed and solidification rate. Calculations for alloy with density increasing as in Figure 13a. From Reference 13.

ingot in Figure 13. In Figure 17, $\Delta C/C_o$ is plotted vs. $\omega^2$ where $\Delta C$ is the composition of Pb at the ingot center minus the composition at the surface, and $C_o$ is the overall ingot composition.

The effect of centrifugal force is to decrease $\Delta C/C_o$ at all solidification rates, but the effect of rotation decreases as the solidification rate increases. For example, at a solidification rate as high as $5.6 \times 10^{-1}$ mm/sec, macrosegregation across the ingot is very slight, and rotation has hardly any effect on the macrosegregation. Note that, although the solidification rate strongly affects macrosegregation, the optimum rotational speed for minimizing macrosegregation is independent of the solidification rate; $\omega$ = 12 rad/sec to achieve $\Delta C$ = 0 in this ingot at all solidification rates.

Figures 13, 15 and 16 illustrate not only the effect of rotation but also effects of density variation on flow of interdendritic liquid and macrosegregation. Specifically for no rotation, contrast Figures 13a, 15a, and 16a. Since interdendritic liquid density depends upon alloy composition, this suggests that segregation can be reduced, to some extent, by constitutional adjustments to alloys. For example, Suzuki and Miyamoto (33) and Lagrange et al. (34) found that an increase of silicon in steel promotes the incidence of "A-type" segregations in conventional type ingots, whereas an increase in molybdenum decreases the segregation. During solidification, these elements are enriched in the interdendritic liquid, and the silicon decreases the liquid density whereas molybdenum increases density. Consequently, silicon enhances and molybdenum suppresses the natural convection of interdendritic liquid in the mushy zone, suggesting that the convection of interdendritic liquid is responsible for the "A-type" segregates found in large steel ingots. Based upon minimizing an integral, I, where

$$I = \int_{g_s}^{1} |\Delta \rho_L| \, g_L \, dg_L \qquad (20)$$

and $|\Delta \rho_L|$ is the absolute value of the difference between the density of the liquid alloy at the liquidus and the density of the enriched interdendritic liquid at some location within the mushy zone, calculations have been carried out to show that by adjusting alloy composition, segregation can be controlled (32). To prevent completely natural convection within the mushy zone, $\Delta \rho_L$ must be zero for all values of $g_L$. In general, this cannot be achieved, but for a multicomponent low alloy steel, it is shown that I can be minimized sufficiently, by alloy adjustment, to eliminate

"A-type" segregates (32). It is suggested, therefore, that alloy design for production of ESR ingots should include calculations which predict macrosegregation for various alloy adjustments.

## MODELLING REFINEMENTS

That the theory of interdendritic liquid flow is applicable to understanding and controlling macrosegregation in ESR ingots is evident. However, there are several refinements which require further study and which should be possibly incorporated into the model. These include: (1) a more complete statement of the energy equation applicable to the mushy zone, (2) experimental work to better evaluate permeability in solidifying alloys, (3) the effect of bulk liquid convection, and (4) the effect of the electromagnetic force field in ESR ingot processing on the flow of interdendritic liquid.

### Energy Equation in the Mushy Zone

Flow of interdendritic liquid within a solidifying alloy can be described as slow-creeping flow; therefore, during solidification of alloys, it is often assumed that the effect of the convection of interdendritic liquid on the temperature distribution is negligible. Thus, the energy equation is written as

$$\rho C_p \frac{\partial T}{\partial t} = \nabla \cdot (k \nabla T) \qquad (21)$$

in which k is thermal conductivity, $\rho$ is density and $C_p$ is the specific heat. When applied to the mushy zone, the thermal conductivity is taken as an average value between solid and liquid phases or the two phases can be approximated as arranged in series or parallel (35). The density of the solid-liquid mixture can be computed according to the relative amounts of solid and liquid but usually the difference in density between liquid and solid phases is only about 3 - 6 percent and so an exact evaluation is not important. The specific heat must include the heat of fusion and several methods of accounting for the heat of fusion have been used. The most realistic is to use the enthalpy, H, of the total alloy, made up of both solid and liquid enthalpies, as a function of temperature during solidification and compute $C_p$ as $C_p = dH/dT$. A similar method is to compute an enthalpy field within a solidifying ingot or casting and then relate enthalpy to temperature.

In large ingots, Equation (21) might not adequately represent the energy equation because the convection of interdendritic liquid should not be ignored (even in the absence of channeled flow). A complete energy balance for a unit element should take into account solidification shrinkage, interdendritic convection, conduction, and generation of heat due to the heat of fusion of the solidifying solid. The temperature within the element is uniform and equal for both solid and liquid phases and varies only differentially across the element. A more complete energy balance, in lieu of Equation (21), is

$$(1-\beta g_L) \frac{\partial T}{\partial t} = \alpha \nabla^2 T - \frac{H_f}{C_p} \frac{\partial g_L}{\partial t} - (1-\beta) g_L \vec{v} \cdot \nabla T \qquad (22)$$

Here, $\alpha$ is the thermal diffusivity, and $H_f$ is the heat of fusion.

The applicability of Equation (22) is shown by examination of Figures 18 and 19. Results of calculations of interdendritic liquid velocity are shown in Figure 18 for a unidirectional steel ingot solidified horizontally from left-to-right. Although flow of interdendritic liquid is only on the order of $10^{-2}$ mm/s, this flow is enough to influence the isotherms within the mushy zone (Figure 19). In Figure 19, the positions of the solidus and liquidus isotherms were specified as a function of time for the purpose of calculations, but the deflections of the other isotherms indicate that the more complete form of the energy balance, Equation (22), should be used in simulating solidification of large ingots rather than Equation (21). Equation (21) does not take into account the convection of interdendritic liquid and its corresponding effect on the distribution of fraction solid within the mushy zone.

## Permeability

Results of calculations given in this paper assume that permeability, K, is independent of orientation. We recognize that this might not be valid but the sparcity of data for permeability in dendritic networks of partially solidified alloys does not justify a more refined comparison between calculation and experiment at this time.

Physically, interdendritic flow is pictured as taking place in long tortuous channels through the porous solid, and to account for the fact that the channels are neither straight nor smooth, a "tortuosity factor," $\tau$, is applied. This model is referred to by Mehrabian et al. (18); with this model the permeability, K, is

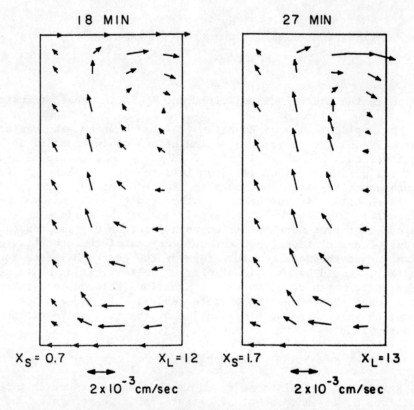

Figure 18: Velocity distribution in the mushy zone of a multi-component low alloy steel. Dimensions are in cm. From Reference 32.

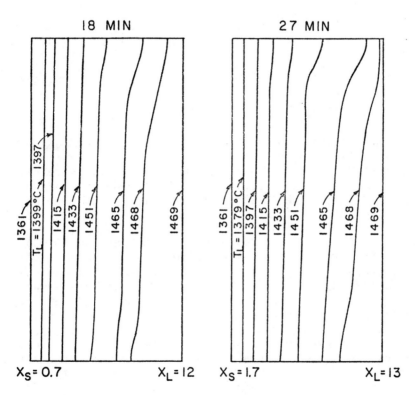

Figure 19: Temperature distribution in the mushy zone shown in Figure 18. Dimensions are in cm. From Reference 32.

$$K = \frac{g_L^2}{8\pi\eta\tau^3} \qquad (23)$$

where $\eta$ is the number of flow channels per unit area normal to flow, and $\tau$ is the tortuosity factor. For a given dendrite arm spacing, if $\eta$ and $\tau$ are constants, the permeability is related to volume fraction liquid by

$$K = \gamma g_L^2 \qquad (24)$$

where

$$\gamma = (8\pi\eta\tau^3)^{-1} .$$

Piwonka and Flemings (36) report values of permeability measured in Al-4% Cu alloy which obey Equation (24) for weight fraction liquid up to about 0.35 with $\gamma \simeq 6 \times 10^{-9} cm^2$. Using a somewhat different experimental technique, Apelian et al. (37) measured permeability in porous dendritic networks of Al-4% Si alloys and also observed that Equation (24) is valid for volume fraction liquid up to 0.37 with $\gamma \simeq 9 \times 10^{-9} cm^2$ and $\gamma \simeq 3 \times 10^{-9} cm^2$ for nongrain-refined and grain-refined structures, respectively. In these works (36,37), permeability was not rigorously determined as a function of dendrite arm spacing. Apelian et al. (37) report approximate secondary dendrite arm spacings of 280 microns (nongrain-refined) and 150 microns (grain refined) measured on two samples only. Piwonka and Flemings (36) do not report dendrite arm spacings, but from examination of micrographs in Piwonka's thesis (38) it appears that the secondary dendrite arm spacing is about 140 microns. The result of these measurements of permeability in aluminum alloys are shown in Figure 20. For $g_L > 0.35$, the slope of the solid curve changes to a value much greater than 2 so that $\gamma$ in Equation (24) is not constant. This can be interpreted to indicate that the number of channels available for flow ($\eta$) or tortuosity ($\tau$) or both are not constant when $g_L > 0.35$. Also, one would expect in the range where $\gamma$ is constant that permeability increase with increasing dendrite arm spacing.

Streat and Weinberg (39) measured permeability in partially solidified Pb-20% Sn at 193°C. In their experiments, volume fraction liquid was held constant ($g_L = 0.19$), and dendrite arm spacing was varied. Using average primary arm spacings ($\bar{\lambda}$), and assuming that

$$n = (\bar{\lambda})^{-2} \qquad (25)$$

they tested measured values of K against Equation (23). As predicted by Equations (23) and (25), permeability varied directly with $\bar{\lambda}^2$ and the tortuosity factor was 4.6.

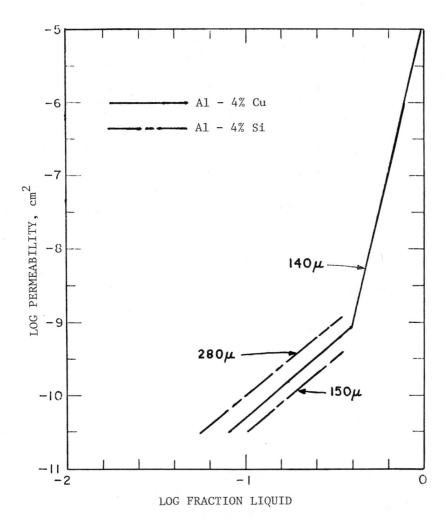

Figure 20: Permeability of aluminum alloys. Numbers on curves are approximate secondary arm spacings; results for Al-4% Cu are from References 36 and 38 and those for Al-4% Si are from Reference 37.

Permeability also varied, qualitatively, as expected with secondary arm spacing, but no simple relationship such as Equation (25) could be readily applied, so primary spacings were considered as the more useful parameter for characterizing the structure. However, one could argue that since secondary arm spacing relates solely to cooling rate during solidification, it is better to characterize structures by secondary arm spacings even though a simple relationship between permeability and secondary arm spacing apparently does not exist. The results of Streat and Weinberg (39) are given in Figure 21 in terms of secondary dendrite arm spacings.

Also shown in Figure 21 are results from Figure 20 at a fraction liquid of 0.19. A number of other values are also represented; these are values of permeability estimated by matching calculated segregation profiles to measured profiles in which permeability was assumed to vary according to Equation 24. It is apparent that there is no correlation between the various sets of data, and the "state of the art" in measuring and/or predicting permeability in partially solidified alloys is not sufficiently advanced to accurately predict values. To sum up, we know that permeability increases with increasing fraction liquid and with increasing dendrite arm spacings and appears also to be a function of alloy; but, as yet, there is no one relationship which takes all of these factors into account quantitatively.

## Electromagnetic Force Field

In an ESR ingot, if the effect of the electromagnetic force field is considered, then Equation (2) takes on the more general form:

$$\vec{v} = -\frac{K}{\mu g_L} (\nabla P + \rho_L \vec{g} + \vec{F}_e) \qquad (26)$$

where $\vec{F}_e$ is the body force vector, due to the electromagnetic force field. Dilawari and Szekely (40) give the necessary expressions to calculate $F_e$ in interacting electric and magnetic fields and solve for the velocity field in the slag phases and the metal pool of cylindrical ESR units. The electrical current is strongest and diverges the most near the electrode in the slag phase and to a lesser extent in the liquid metal pool; therefore $\vec{F}_e$ produces strong convection in the slag and influences convection in the liquid pool to a lesser extent. Since the lines of current are more nearly parallel in the mushy zone, then the strength of $\vec{F}_e$ is expected to be weaker than the force due to gravity and probably can be ignored when calculating flow of interdendritic liquid. However, this assumption should be verified; this remains to be done.

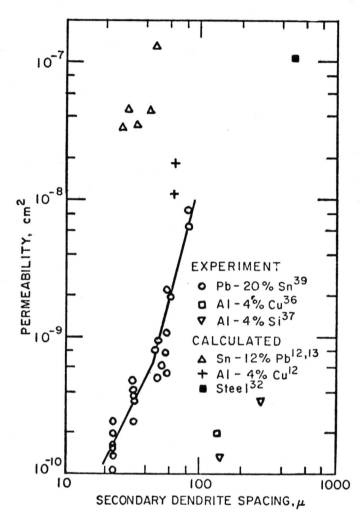

Figure 21: Permeability of partially solidified alloys ($g_L$ = 0.19). Shown are values experimentally measured and values obtained by matching calculated segregation to measured segregation.

As discussed earlier in this paper, the electromagnetic force field contributes to the overall convection of the bulk liquid (23,40) and, therefore, possibly has an indirect effect on the flow of interdendritic liquid in the mushy zone of an ESR ingot.

## ACKNOWLEDGEMENTS

Some of the research reported herein was sponsored by the Army Materials and Mechanics Research Center, Watertown, Massachusetts, under Contract No. DAAG46-74-C-0120. Continuing work has been sponsored by the National Science Foundation as part of the U.S.A.-U.S.S.R. Cooperative Program in Electrometallurgy under Grant No. DMR 76-03682.

Many coworkers contributed significantly to the underlying research and deserve our sincere gratitude. They are S. D. Ridder, F. C. Reyes and S. Chakravorty at the University of Illinois-Urbana, J. H. Chen at Cabot Corporation, and T. Fujii and R. Furlong at M.I.T. Sindo Kou deserves special thanks. He has done research on macrosegregation both at M.I.T. and the University of Illinois and is coauthor of previous papers and reports on macrosegregation.

## References

1. S. G. Arwidson, "Technico-Economic Appraisal of a Large Tonnage ESR Installation," in Electroslag Refining, Iron and Steel Institute, London, 1973, pp. 157-162.

2. H. Fredriksson and O. Jarleborg, J. of Metals, September, 1971, p. 32.

3. J. O. Ward and R. C. Hambleton, "Production Experience of Electroflux Remelting of Nickel-Base Superalloys," in Electroslag Refining, Iron and Steel Institute, London, 1973, pp. 80-88.

4. M. Basaran, T. Z. Kattamis, R. Mehrabian, and M. C. Flemings, "A Study of the Heat and Fluid Flow in Electroslag Remelting," Army Materials and Mechanics Research Center, Contract No. DAAG 46-73-C-0088, April, 1974.

5. Yu M. Mironov and M. M. Klyuev, Russian Metallurgy, 1968, No. 2, pp. 72-76.

6. M. A. Maulvault, "Temperature and Heat Flow in the Electroslag Remelting Process," Ph.D. Thesis, Department of Materials Science and Engineering, M.I.T., 1971.

7. A. S. Ballantyne and A. Mitchell, Ironmaking and Steelmaking, 1977, No. 4, pp. 222-239.

8. R. C. Sun, J. W. Pridgeon, "Predicting Pool Shapes in a Laboratory Electroslag Remelting Process," Union Carbide Corporation, Materials System Division, Technology Department, Kokomo, Indiana, September, 1969, Report No. 7649.

9. A Mitchell, S. Joshi, Met. Trans., V. 4, 1973, pp. 631-642.

10. B. Z. Paton et al., Calculation of Temperature Fields in Plate Ingots and in Ingot-Slabs of ESR," in <u>Fifth International Symposium on ESR Process</u>, Carnegie-Mellon University, Pittsburgh, October, 1974, p. 323.

11. S. D. Ridder, F. C. Reyes, S. Chakravorty, R. Mehrabian, J. D. Nauman, J. H. Chen and H. J. Klein, "Steady State Segregation and Heat Flow in ESR," to be published in Met. Trans. B., 1978.

12. S. Kou, D. R. Poirier, and M. C. Flemings, "Macrosegregation in Electroslag Remelted Ingots," in <u>Proceedings of the Electric Furnace Conference</u>, Iron and Steel Society of AIME, December, 1977.

13. S. Kou, D. R. Poirier and M. C. Flemings, "Macrosegregation in Rotated Remelted Ingots," to be published in Met. Trans. B, 1978.

14. M. C. Flemings, G. E. Nereo, Trans. Met. Soc., AIME, V. 239, 1967, pp. 1449-1461.

15. M. D. Flemings, R. Mehrabian and G. E. Nereo, Trans. Met. Soc., AIME, V. 242, 1968, pp. 41-49.

16. M. C. Flemings and G. E. Nereo, Trans. Met. Soc., AIME, V. 242, 1968, pp. 50-55.

17. R. Mehrabian and M. C. Flemings, Trans. Met. Soc., AIME, V. 245, 1969, p. 2347.

18. R. Mehrabian, M. A. Keane and M. C. Flemings, Met. Trans., V. 1, 1970, p. 1209.

19. R. Mehrabian, M. A. Keane and M. C. Flemings, Met. Trans. V. 1, 1970, p. 3238.

20. M. Keane, Sc.D. Dissertation, Massachusetts Institute of Technology, 1973.

21. R. Mehrabian, J. J. Burke, M. C. Flemings and A. E. Gorum, Eds., <u>Solidification Technology</u>, Brook Hill Publishing Company, Chestnut Hill, Mass., 1974.

22. M. C. Flemings, Scandinavian J. of Metallurgy, V. 5, 1976, p. 1.

23. A. H. Dilawari and J. Szekely, Met. Trans. B, V. 9B, 1978, pp. 77-87.

24. E. Scheil, Metallforsch., V. 2, 1947, p. 69.

25. J. S. Kirkaldy and W. V. Youdelis, Trans. AIME, V. 212, 1958, p. 833.

26. S. Kou, "Macrosegregation in Electroslag Remelted Ingots," Ph.D. Thesis, Department of Materials Science and Engineering, Massachusetts Institute of Technology, 1978.

27. D. R. Poirier, S. Kou, T. Fujii and M. C. Flemings, "Electroslag Remelting," Contract No. DAAG46-74-C-0120, Army Materials and Mechanics Research Center, Watertown, Mass., AMMRC TR78-28, June, 1978.

28. P. O. Mellberg and H. Sandberg, Scand. J. Metallurgy, V. 2, 1973, pp. 83-86.

29. R. H. Frost, "Solidification of Electroslag Remelted Low Alloy Steel Ingots," Army Materials and Mechanics Research Center, Watertown, Mass., AMMRC TR77-20, October, 1977.

30. R. J. McDonald and J. D. Hunt, TMS-AIME, V. 245, 1969, p. 1993.

31. S. M. Copley, A. F. Giamei, S. M. Johnson and J. F. Hornbecker, Met. Trans., V. 1, 1970, p. 2193.

32. T. Fujii, D. R. Poirier and M. C. Flemings, "Macrosegregation in a Multicomponent Low Alloy Steel," submitted for publication to Met. Trans., June, 1978.

33. K. Suzuki and T. Miyamoto, Japan Society for the Promotion of Sciences, Nineteenth Committee, 1972, No. 9478.

34. J. Lagrange, J. Delorme and P. Bocquet, "Influence du Mode d'Elaboration sur la Segregation des Gros Lingots de Forge," 20eme Colloque de Metallurgie, Saclay, Juin, 1977.

35. T. W. Caldwell, A. J. Campagna, M. C. Flemings and R. Mehrabian, Met. Trans. B, V. 8B, 1977, pp. 261-270.

36. T. S. Piwonka, M. C. Flemings, "Pore Formation in Solidification," Trans. Met. Soc. AIME, V. 236, 1966, pp. 1157-1165.

37. D. Apelian, M. C. Flemings, R. Mehrabian, "Specific Permeability of Partially Solidified Networks of Al-Si Alloys," Met. Trans., V. 5, 1974, pp. 2533-2537.

38. T. S. Piwonka, Sc.D. Thesis, Department of Metallurgy, Massachusetts Institute of Technology, 1964.

39. N. Streat and F. Weinberg, Met. Trans. B, V. 7B, 1976, p. 417.

40. A. H. Dilawari and J. Szekely, Met. Trans. B, V. 8B, 1977, pp. 227-236.

# THE ANALYSIS OF MAGNETOHYDRODYNAMICS AND PLASMA DYNAMICS IN METALS PROCESSING OPERATIONS

C. W. Chang, J. Szekely and T. W. Eagar

Department of Materials Science and Engineering
Massachusetts Institute of Technology
Cambridge, Massachusetts  02139

The principles of magnetohydrodynamics and plasma dynamics are reviewed in terms of their applications to fluid flow behavior in metals processing operations. The mathematical formulation of the problem is described in detail, while specific examples include induction stirring, electroslag welding, electric arc melting, and arc welding. It is shown that the calculated predictions of flow behavior in these operations correspond closely to the experimental data and, hence, the procedure outlined provides a useful new tool in understanding metals processing systems.

## 1  INTRODUCTION

There are numerous processing operations involving super-alloys and other strategic materials, where the electromagnetic force field plays an important role. Typical examples include electroslag refining, electroslag welding, plasma arc melting, induction stirring, and the like.

A common feature of the systems is that the thermal energy is generated within the system by means of an applied or induced electric current. In general, this current will interact with the generated magnetic field, which establishes an electromagnetic force field. This electromagnetic force field produces motion in the melt, which may play a significant role in determining the process characteristics.

Our purpose in this paper is to present a brief review of the fundamentals that govern the electromagnetically driven flows and then to illustrate the application of these concepts through specific examples, which will include induction stirring, electroslag welding, and plasma technology.

## 2 MAGNETOHYDRODYNAMICS AND ITS APPLICATIONS TO LIQUID MELTS

In order to analyze the fluid flow behavior in systems containing either applied or induced electric currents, it is necessary to calculate the electromagnetic force field. This may be done by applying the magnetohydrodynamic (MHD) approximations of Maxwell's equations to the system of interest.

### 2.1 Equations under the MHD Approximation[1,2]

Under the MHD approximation, Maxwell's equations may be written:

(Faraday's law)     $\nabla \times \underline{E} = -\frac{\partial \underline{B}}{\partial t}$     (1)

(Ampere's law)     $\nabla \times \underline{H} = \underline{J}$     (2)

where $\underline{E}$ is the electric field,
$\underline{B}$ is the magnetic flux density,
$\underline{H}$ is the magnetic field intensity,
$\underline{J}$ is the current density,
t is time.

Furthermore, we have that

$$\underline{B} = \mu_0 \underline{H} \qquad (3)$$

where $\mu_0$ is the magnetic permeability of free space.

The current density is given by Ohm's law:

$$\underline{J} = \sigma(\underline{E} + \underline{v} \times \underline{B}) \qquad (4)$$

where $\sigma$ is the electrical conductivity and $\underline{v}$ is the velocity. In many materials processing operations, the $\underline{v} \times \underline{B}$ term of Eq. (4) is small compared to the electric field term and, hence, may be neglected, resulting in considerable simplification of the calculations. Additionally, we have the equation of continuity,

$$\nabla \cdot \underline{v} = 0, \qquad (5)$$

and the time-smoothed equation of motion:

$$\rho(\underline{v} \cdot \underline{\nabla}) v = -\underline{\nabla} p + \underline{\nabla} \cdot \tau + \underline{F}_b \qquad (6)$$

where $\rho$ is the density,
$\nabla p$ is the pressure gradient,
$\tau$ is the stress tensor,
$\underline{F}_b$ is the body force.

In this study, $\underline{F}_b$ is the Lorentz force and may be calculated from:

$$\underline{F}_b = \underline{J} \times \underline{B}. \qquad (7)$$

### 2.2 Principles of Induction Stirring

Faraday's law states that alternating magnetic fields will produce induced currents in molten or solid metals. These induced currents generate both heat, which is useful for the remelting of metal scrap, and electromagnetic forces, which cause stirring and homogenization of the liquid pool. In this paper, the principles of induction stirring will be described.

Let us consider a melt contained in an induction furnace, such as that sketched in Figure 1. Current is supplied through the coils, which, according to Ampere's law, creates a magnetic field. This magnetic field interacts with the molten metal in accordance with Faraday's and Ohm's laws to produce an induced electric current, J. This current in turn interacts with the induced magnetic field, $\underline{B}$, to produce an electromagnetic force field (Equation (7)). As noted previously, it is this force field which causes stirring of the liquid pool. The general nature of the solution to these equations will now be discussed.[3,4]

<u>Ratio of conduction and induction current</u>. By dimensional arguments it may be shown from Equation (1) that the magnitude of the electric field is the order of:*

$$O(\underline{E}) \sim B_0 fL \qquad (8)$$

where f is characteristic frequency,
L is characteristic length,
$B_0$ is characteristic magnetic flux density.

Similarly, from Ohm's law and Equation (8) above, the ratio of the conduction current to the induction current is of the order:

---

*In this paper, the symbol $O(\underline{E})$ is used to denote "the order of."

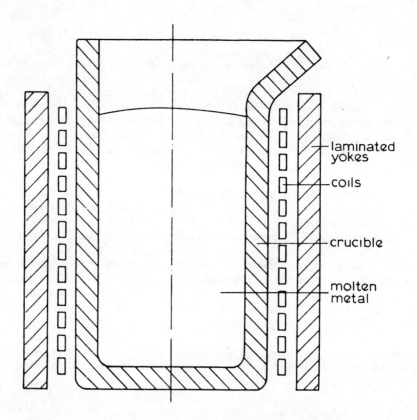

Figure 1: Sketch of an induction furnace.

$$O\left\{\frac{\underset{\sim}{E}}{\underset{\sim}{v} \times \underset{\sim}{B}}\right\} \sim \frac{fL}{u_o} \tag{9}$$

where $u_o$ is a characteristic velocity of the system.

In a typical high frequency induction furnace, the velocity is small compared to the product of the frequency and the characteristic length, that is:

$$fL \gg u_o, \tag{10}$$

hence, for induction stirring, it is reasonable to neglect the induction current in Ohm's law. That is,

$$\underset{\sim}{J} \stackrel{\sim}{=} \sigma E, \tag{11}$$

which, as noted previously, permits a considerable simplification of the calculation.

Relationship between velocity and coil current. An important objective in the modelling of flow fields in induction furnaces is to relate the coil current to the velocity. If the inertial forces dominate, then as a first approximation, we may neglect the first two terms on the right-hand side of Equation (6). Considering the order of magnitude of the quantities in the remaining terms, we have

$$O\{\rho \ (\underset{\sim}{v} \cdot \underset{\sim}{\nabla}) \underset{\sim}{v}\} \frac{\rho u_o^2}{L}, \tag{12}$$

$$O\{\underset{\sim}{J} \times \underset{\sim}{B}\} \sim \sigma B_o^2 fL. \tag{13}$$

On taking

$$B_o \sim J_o \mu_o L \tag{14}$$

where $J_o$ is the coil current, and combining Equations (12) and (13), we have

$$\frac{\rho u_o^2}{L} \sim \sigma fL^3 J_o^2 \mu_o^2, \tag{15}$$

i.e.,

$$u_o \sim J_o \sqrt{\frac{\sigma f}{\rho}} L^2 \mu_o. \tag{16}$$

This linear relationship between coil current and the velocity has been verified experimentally by Dragunkina and Tir[5] as is shown in Figure 2.

**Nature of stirring.** Using Ampere's law, the Lorentz force can be rewritten as:

$$\underset{\sim}{J} \times \underset{\sim}{B} = -\underset{\sim}{\nabla}(\frac{1}{2\mu_o} B^2) + \frac{1}{\mu_o} \underset{\sim}{B} \cdot \underset{\sim}{\nabla}\underset{\sim}{B}. \tag{17}$$

Taking the curl of the right-hand side of Equation (17) we note that:

$$\underset{\sim}{\nabla} \times (-\underset{\sim}{\nabla}(\frac{1}{2\mu_o} B^2)) = 0, \tag{18}$$

while, in general,

$$\underset{\sim}{\nabla} \times \frac{1}{\mu_o} \underset{\sim}{B} \cdot \underset{\sim}{\nabla}\underset{\sim}{B} \neq 0. \tag{19}$$

It follows that the first term on the right-hand side of Equation (17) can do no work on a circulating body of molten metal, thus the second term, i.e., $\frac{1}{\mu_o} \underset{\sim}{B} \cdot \underset{\sim}{\nabla}\underset{\sim}{B}$, is responsible for the work done on the circulating body of metal and, hence, causes the stirring.

**Computed and measured results.** The foregoing principles have been applied to several induction stirred systems with considerable success.

i) Melt velocities in a low melting alloy system. In one study,[6] local velocities and magnetic field strength measurements were made on a 250mm diameter stainless steel crucible containing Wood's metal. A sketch of the apparatus is shown in Figure 3. The experimental velocity measurements are given in Figure 4. Subsequent numerical calculations on this system using the above described MHD principles provided the velocity mapping shown in Figure 5. It will be noted that the computed solution is in reasonable agreement with the experimental data. Furthermore, the computation technique provides a much more comprehensive mapping of the flow behavior of the melt than could reasonably be obtained from experimental techniques alone.

ii) Melt surface velocities in a 30,000 lb. inductively

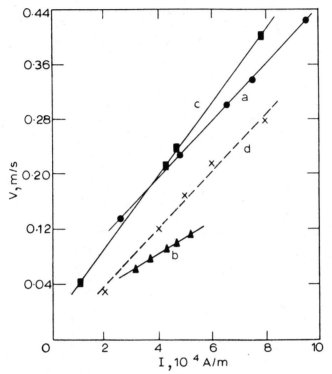

*a* f=50 Hz with cover; *b* f=2500 Hz with cover; *c* f=50 Hz without cover; *d* f=1060 Hz

Figure 2: Comparison of velocities of aluminum at axis of induction furnace at frequencies of 50 and 2500 Hz.[4,5]

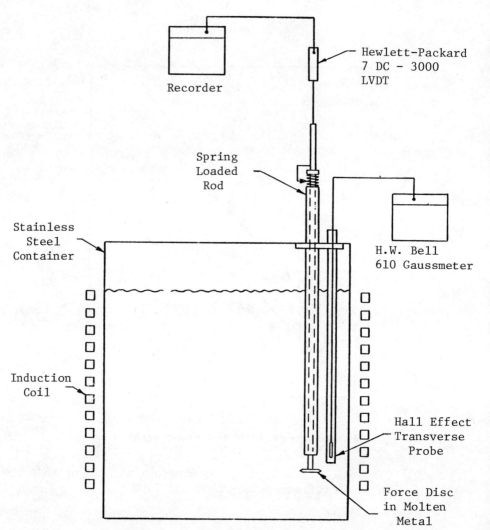

Figure 3: Schematic sketch of the apparatus used in studying induction stirring of low melting-alloys.

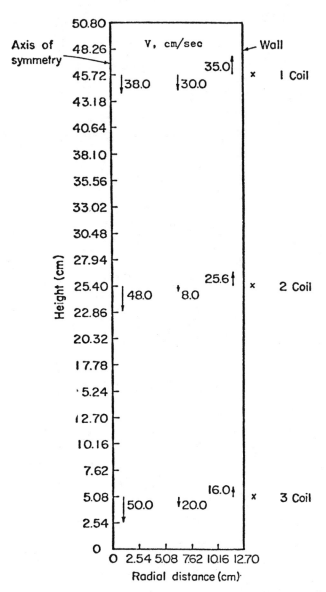

Figure 4: The experimentally measured velocity field found in low alloy induction melts.[6]

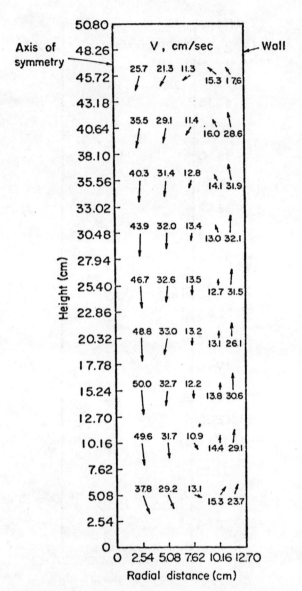

Figure 5: The computed velocity field calculated for low alloy induction melts.[6]

stirred melt. Another experimental verification of the power of the computational technique described herein, has been given in a recent paper[7] on vacuum induction stirring of nickel base superalloys. The experimental set up is given in Figure 6, which shows a TV camera used to follow the melt surface velocities in a large crucible. A comparison of the measured and computed average surface velocities is given in Table I. It is seen that the agreement between experiment and theory is remarkable; and it should again be noted that the computation technique provides significantly more information about the system than can be readily obtained experimentally.

### 2.3 Principles of Electroslag Welding

In terms of weld metal deposition rates, electroslag welding (ESW) is the most efficient process for joining thick plates, and as such is potentially attractive in the fabrication of bridges, ships, and pressure vessels. The process as sketched in Figure 7 consists of a rectangular cavity approximately one to one and one-half inches thick, bounded on two sides by the plates to be joined and on the other two sides by dams or shoes which are typically made of water-cooled copper. A small quantity of powdered flux is placed in the bottom of the cavity and a consumable welding electrode is guided to the bottom where a momentary arc causes the flux to melt. Once the molten slag has formed, the ESW process has begun. Heat is generated by the resistance of the welding current passing through the molten slag to the base plates. The current induces a self-magnetic field, which, due to the divergence of the current path from the electrode wire to the plate, is spatially non-uniform. This magnetic field interacting with the welding current produces a spatially varying electromagnetic force field (Equation (7)) which, coupled with a thermal buoyancy force field, is responsible for the vigorous slag and metal flow observed in this process. Due to the small dimensions and high temperatures involved in the process, experimental measurement of the flow behavior is extremely difficult; however, utilizing the analytical techniques described previously, the process may be modelled mathematically.[8,9]

Computed results. The ESW system was modelled in two dimensions in order to simplify the calculations. The rectangular cavity was replaced by one of cylindrical geometry with a consumable electrode located along the central axis. Due to the large thermal gradients inherent in the process, account was made of the thermal buoyancy forces. Figures 8 and 9 give the streamline pattern and velocity vector mapping of one-half the system.[8,9] It will be noted in Figure 9 that the maximum calculated velocities are on the order of 40 cm/sec, which compares very favorably with the experimental measurements of Patchett and Milner[10] of 50 - 100

## TABLE I

### Measured and Computed Average Surface Velocities

|  | Measured | Computed |
|---|---|---|
| Average Surface Velocity, cm/sec (500 KW; 2500 A) | 18.19 ± 3.55 | 17.6 |
| Average Surface Velocity, cm/sec (2000 KW; 5000 A) | 39.56 ± 7.59 | 32.7 |
| Ratio of Average Surface Velocity, (2000 KW/500 KW) | 2.17 | 1.86 |

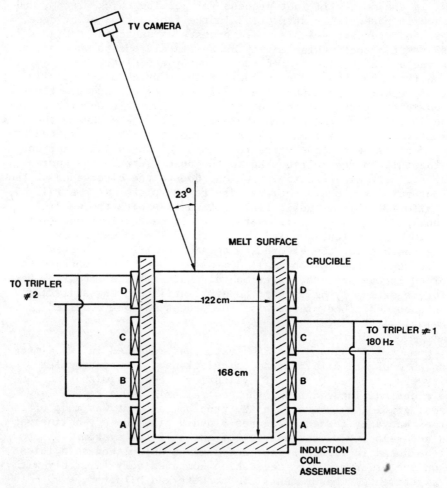

Figure 6: Sketch of the crucible and coil configuration of a 30,000 lb. vacuum induction melter.

# MAGNETOHYDRODYNAMICS AND PLASMA DYNAMICS

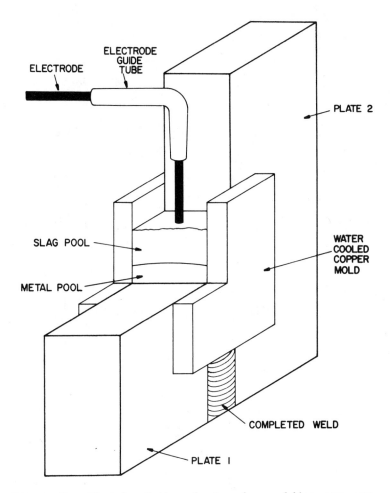

Figure 7: Sketch of the electroslag welding process.

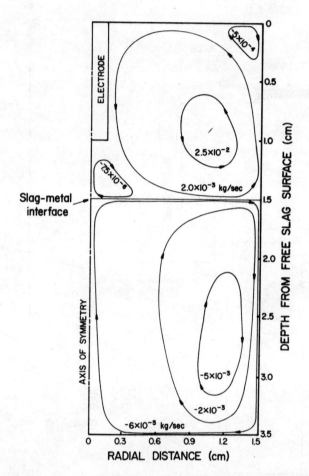

Figure 8: Computed streamline pattern for a cylindrical electroslag welding system with heat input of 4.3 Kcal/sec.[8,9]

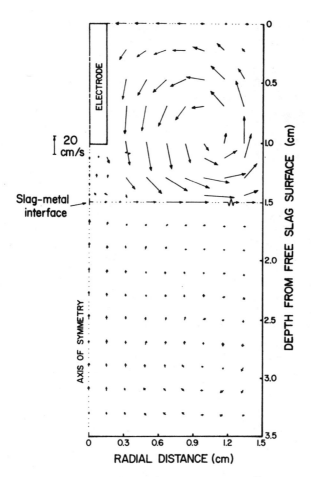

Figure 9: Computed map of the velocity vector for a cyclindrical electroslag welding system with heat input of 4.3 Kcal/sec. [8,9]

cm/sec in aluminum alloys. In addition, the recirculating loops as seen in Figure 8 tend to explain the chemical heterogeneities found in the solidified slags by the same authors.[10] Hence, it is seen that the analytical techniques described herein, provide insights and understandings of the ESW process which would be difficult, if not impossible, to obtain experimentally.

## 3 PLASMA DYNAMICS AND ITS APPLICATIONS TO ARCS USED IN METALS PROCESSING

The electric arc is by far the most important application of plasma physics used in industry, and has a very broad application including electric arc furnaces, arc welding, and plasma generators, to cite but a few representative examples. The definition of the arc, in itself, is a problem since the same word has been used to describe a number of electrical discharges looking more or less alike, but lacking a common physical mechanism. In general, plasma physicists seem to agree that the arc may be defined in terms of current and voltage drop only.

The arc exists as the result of the passage of electricity through a gas between two electrodes, and it can be formed only under conditions where the continuity of current can be maintained by processes occurring within the arc itself. This can be most easily demonstrated for the simple thermionic arc, in which a carbon or tungsten cathode is able to supply the required electrons as a result of heating to a temperature at which adequate thermionic emission can take place. In such an arc electrons are emitted from the cathode and become accelerated under the applied field to gain energy, part of which is then lost by collision with the gas molecules, so that the gas is raised to a high temperature at which it becomes thermally ionized. The positive ions so formed travel through the arc plasma in the opposite direction to that of the electrons and eventually lose a large proportion of their energy in collision with the cathode, which is thereby maintained at the temperature necessary for the emission of the required number of electrons.

For all arcs, the conducting gas between the electrodes has a high temperature and high luminosity. The conducting arc column acts as a normal electrical conductor in the presence of a magnetic field, although anomalous electromagnetic effects have been observed at the cathode. Because of its gaseous nature, the arc is easily influenced by gas flow. However, the spatial stability of arcs is greatly dependent on the nature of the cathode material. Thus, with typically refractory cathodes, e.g., carbon, molybdenum, and tungsten, the cathode temperature is high and the arc is relatively stable. With low-melting-point cathodes, e.g., copper and mercury, the cathode termination is a highly mobile and concentrated spot which moves constantly over the cathode surface in an irregular fashion.

Figure 10 is a sketch of typical characteristics of the free burning arc. The luminous plasma region approximates a bell shape in which cylindrical symmetry is assumed to be maintained. There are three regions of markedly different behavior in the arc, the cathode-fall region, the plasma column or positive column, and the anode-fall region.

Since the positive column satisfies the definition of a high-pressure thermal plasma, a brief introduction of the application of plasma physics to the electric arc will be presented.

From a theoretical point of view, the basic description of a plasma lies in the kinetic theory of matter. One defines a function of position, velocity, and time, $f(\underline{r},\underline{v},t)$, such that $f\,d\underline{r}\,d\underline{v}$ is the probability of finding particles within the six-dimensional volume element $d\underline{r}\,d\underline{v}$, centered at the point $(\underline{r},\underline{v})$ in coordinate and velocity space. Observable properties of the plasma can then be obtained from this function, known as the distribution function, by taking various velocity moments of f.

The equation determining the distribution function is called the kinetic equation:

$$\frac{\partial f}{\partial t} + \underline{v} \cdot \nabla_r f + \frac{\underline{F}}{m} \cdot \nabla_v f = \left(\frac{\partial f}{\partial t}\right)_c \tag{20}$$

where $\nabla_r$ is the gradient with respect to position,
$\nabla_v$ is the gradient with respect to velocity, and
the term $(\partial f/\partial t)_c$ is due to collision.
This kinetic equation is well known as the Boltzmann equation.

In principle, it is easy to deal with the average field due to the charged particles, since this is given by Poisson's equation:

$$\nabla \cdot \underline{E}\,(\underline{r},t) = \frac{1}{\varepsilon_o} \sum_i q_i \int f_i\,(\underline{r},\underline{v},t)\,d\underline{v} \tag{21}$$

where $\varepsilon_o$ is the permittivity of free space, q is the electric charge and the index has been introduced to denote particle type.

The field $\underline{E}$ is often referred to as the self-consistent field since it depends on f and its evaluation therefore requires a self-consistent solution of the kinetic equation (for each species of particles) together with the Maxwell's equations for the electric and magnetic fields;

$$\nabla \times \underline{E} = -\frac{\partial \underline{B}}{\partial t} \tag{22}$$

$$\nabla \times \underline{B} = -\varepsilon_o \frac{\partial \underline{E}}{\partial t} + \underline{J}, \tag{23}$$

$$\nabla \cdot \underline{B} = 0 \tag{24}$$

where the current density is

$$\underline{J}(\underline{r},t) = \sum_i q_i \int \underline{v}\, f_i(\underline{r},\underline{v},t)\, d\underline{v}. \tag{25}$$

The particles are subjected to a Lorentz force

$$\underline{F} = q(\underline{E} + \underline{v} \times \underline{B}). \tag{26}$$

For a collisionless plasma, we have

$$\frac{\partial f}{\partial t} + \underline{v} \cdot \nabla_{\underline{r}} f + \frac{q}{m}(\underline{E} + \underline{v} \times \underline{B}) \cdot \nabla_{\underline{v}} f = 0, \tag{27}$$

which is called the Vlasov equation.

The governing equations to describe the fluid model of the plasma can be obtained from the moment equations. The moment equations are obtained from Equation (27) by multiplying the various functions of velocity $\psi(\underline{v})$, and integrating over velocity space. Essentially, $\psi(\underline{v})$ takes the values 1, $\underline{v}$, $\underline{v}\underline{v}$; thus, giving rise to the zero-order, first-order, and second-order moment equations respectively. These equations are usually called equation of continuity, Navier-Stokes equation, and energy equation, respectively. The detailed derivations are omitted and can be found in standard textbooks.[11-13]

Two sets of equations can be written for an electron-ion plasma. In order to simplify the inherent complexity of the problem, we can define some quantities based on average values of a mixture of electrons and ions. After some algebraic manipulations as described in

# MAGNETOHYDRODYNAMICS AND PLASMA DYNAMICS

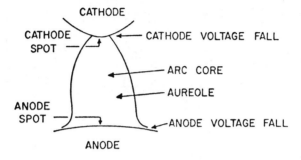

Figure 10: Sketch of a free burning arc.

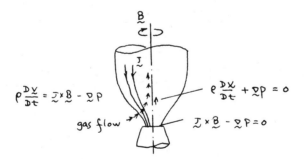

Figure 11: Plasma flow field for a free burning arc.

the textbooks,[11-13] it is possible to derive the single fluid MHD equations as has been described previously in Equations (1) through (4). However, due to the compressibility of the gas, the equation of continuity now takes the form,

$$\nabla \cdot (\rho \underline{v}) = 0, \tag{28}$$

while the equation of motion becomes

$$\nabla \cdot (\rho \underline{v}\underline{v}) = -\nabla P + \nabla \cdot \tau + \underline{J} \times \underline{B}, \tag{29}$$

and the equation of state for an ideal gas is written

$$P = \rho RT \tag{30}$$

where R is the gas content.

### 3.1 Principles of the Electric Arc

<u>Order of magnitude analysis</u>. Let us consider the model shown in Figure 11. By dimensional arguments it may be shown from Equations (2) and (3) that

$$O(\underline{B}) \sim \mu_o L J_o,$$

which is identical to Equation (14).

An important objective in the modelling of flow fields in the electric arc is to relate the applied current to the velocity. As a starting point, let us consider the Navier-Stokes equation (Equation (29)), which, under steady state conditions with constant density and viscosity, becomes:

$$\rho (\underline{v} \cdot \nabla) \underline{v} = -\nabla P + \mu \nabla^2 \underline{v} + \underline{J} \times \underline{B} \tag{32}$$

If the inertial forces dominate and the viscous forces can be neglected, we may proceed in a manner similar to that described in the section on induction stirring, by considering the order of magnitude of the following quantities of Equation (32):

$$O\{\rho (\underline{v} \cdot \nabla) \underline{v}\} \sim \frac{\rho u_o^2}{L} \tag{12}$$

and $O\{\underset{\sim}{J} \times \underset{\sim}{B}\} \sim J_o^2 L\mu_o$ . (33)

Since the electromagnetic force is balanced by the inertial force, the combination of Equations (12) and (33) gives

$$u_o \sim J_o L \sqrt{\frac{\mu_o}{\rho}} \ .$$ (34)

This approximate linear relationship between the applied current and the velocity is similar to the result predicted by Maecker.[14] That is,

$$u_{max} = \sqrt{\frac{\mu_o}{2\rho}} \ r_c J_o$$ (35)

where $r_c$ is the redius of the cathode.

If the electromagnetic force is dissipated by the viscous force, from Equation (32) we have

$$O\{\mu \nabla^2 \underset{\sim}{v}\} \sim O\{\underset{\sim}{J} \times \underset{\sim}{B}\}$$ (36)

Similarly, using Equations (33) and (36), we can deduce the following result:

$$u_o \sim (\frac{\mu o}{\mu} L^3) J_o^2 \ .$$ (37)

This approximate relationship between the applied current and the velocity may be used to interpret the results proposed by Woods et al.[15] in their study of the velocities in a bounded mercury bath, which is shown in Figure 12.

### 3.2 Application of Plasma Dynamic Principles to Electric Arc Melting

Current work at MIT using these plasma dynamic principles is aimed at understanding arc stability, heat transfer to the melt, and velocity and thermal distributions in electric arc furnaces. The objective of this research is to improve the process control and efficiency. Currently, approximately 20 to 25 percent of the steel made in the United States is being produced in electric arc furnaces, which corresponds to an annual power consumption of 13.5 billion KW hours. Hence, even modest improvements in the process efficiency represent significant savings in energy.

### 3.3 Application of Plasma Dynamic Principles to Arc Welding

Ever since Maecker[14] first described plasma jet behavior in arcs, welding technologists have used this phenomenon to explain numerous arc welding effects such as metal transfer,[16] penetration,[17] and puddle distortion.[18] While several studies have attempted to estimate[19] or measure[15] the magnitude of the fluid flow within the arc, current understanding of this phenomenon is only qualitative. It is expected that the techniques described in this paper will aid in quantifying our understanding of the arc welding process. Furthermore, it is expected that the MHD approach will be useful in calculating the flow behavior in the molten weld pool itself. The implications of this type of study to segregation and defect formation in weldments are likely to be enormous.

### 4: CONCLUSIONS

The analysis of magnetohydrodynamic and plasma dynamic principles in several metals processing systems has been described. It has been shown that the mathematical techniques described herein provide very close agreement with the limited experimental data available. In addition, it is seen that the mathematical technique is capable of providing significantly greater amounts of in-

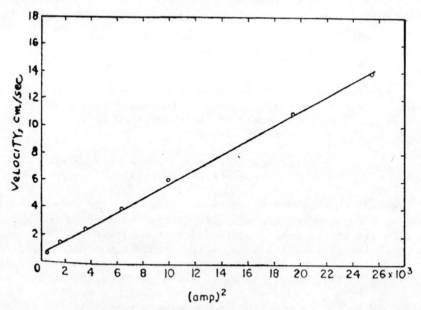

Figure 12: The velocity of the central stream of a mercury bath as a function of applied current.[15]

formation about the process than can be readily obtained experimentally. It is though that the ideas presented will provide important insights into materials processing systems for many years to come.

## ACKNOWLEDGEMENTS

Support of the research presented in this paper has been provided by the National Science Foundation and the United States Department of Energy and is gratefully acknowledged.

## References

1. W. F. Hughes and F. J. Young, The Electromagnetodynamics of Fluids, John Wiley & Sons, New York, 1966.

2. J. A. Shercliff, A Textbook of Magnetohydrodynamics, Pergamon Press, 1965.

3. J. Szekely and C. W. Chang, "Turbulent Electromagnetically Driven Flow in Metals Processing: Part I Formulation," Ironmaking and Steelmaking, 1977, No. 3, p. 190.

4. J. Szekely and C. W. Chang, "Turbulent Electromagnetically Driven Flow in Metals Processing: Part II Practical Applications," Ironmaking and Steelmaking, 1977, No. 3, p. 196.

5. N. I. Dragunkina and L. L. Tir, "Experimental Investigation of Similarity Conditions in the Motion of Molten Metal in an Induction Furnace," Magnetohydrodynamics, 1966, Vol. 2, No. 1, p. 81.

6. J. Szekely, C. W. Chang and R. E. Ryan, "The Measurement and Prediction of the Melt Velocities in a Turbulent, Electromagnetically Driven Recirculating Low Melting Alloy System," Met. Trans. B, 1977, Vol. 8B, p. 333.

7. J. Szekely, C. W. Chang and W. E. Johnson, "Experimental Measurement and Prediction of Melt Surface Velocities in a 30,000 lb. Inductively Stirred Melt," Met. Trans. B, 1977, Vol. 8B, p. 514.

8. A. H. Dilawari, T. W. Eagar and J. Szekely, "An Analysis of Heat and Fluid Flow Phenomena in Electroslag Welding," Welding Journal, Jan. 1978, p. 24-s.

9. A. H. Dilawari, J. Szekely and T. W. Eagar, "Electromagnetically and Thermally Driven Flow Phenomena in Electroslag Welding," Met. Trans. B (in press).

10. B. M. Patchett and D. R. Milner, "Slag-Metal Reactions in the Electroslag Process, " <u>Welding Journal</u>, Oct. 1972, p. 491-s.

11. F. F. Chen, <u>Introduction to Plasma Physics</u>, Plenum Press, New York, 1974.

12. T. J. Boyd and J. Sanderson, Plasma Dynamics, Barnes and Noble, New York, 1970.

13. J. L. Shohet, <u>The Plasma State</u>, Academic Press, New York, 1971.

14. H. Maecker, "Plasmaströmungen in Lichtbögen infolge eigenmagnetischer Kompression," <u>Z. Phys.</u>, 1955, Vol. 141, p. 198.

15. R. A. Woods and D. R. Milner, "Motion in the Weld Pool in Arc Welding," <u>Welding Journal</u>, 1971, Vol. 50, p. 163-s.

16. J. C. Needham, C. J. Cooksey and D. R. Milner, "Metal Transfer in Inert-Gas Shielded-Arc Welding," <u>Brit. Welding Journal</u>, 1960, p. 101.

.17. K. R. Spiller and G. J. MacGregor, "Effect of Electrode Vertex Angle on Fused Weld Geometry in TIG-Welding," in <u>Advances in Welding Processes</u>, Welding Institute, 1970, p. 83.

18. E. Friedman, "Analysis of Weld Puddle Distortion and Its Effect on Penetration," <u>Welding Journal</u>, 1978, Vol. 57, No. 6, p. 161-s.

19. T. B. Reed, "Determination of Streaming Velocity and the Flow of Heat and Mass in High-Current Arcs," <u>J. Appl. Phys.</u>, 1960, Vol. 31, p. 2048.

## Notations

| | |
|---|---|
| $\tilde{B}$ | magnetic flux density |
| $B_o$ | characteristic magnetic flux density |
| $\tilde{E}$ | electric field |
| $\tilde{f}$ | characteristic frequency; distribution function |
| $\tilde{F}$ | force; Lorentz force |
| $\tilde{F}_b$ | body force; Lorentz force |
| $\tilde{H}$ | magnetic field intensity |
| $\tilde{J}$ | current density |
| $\tilde{J}_o$ | current density of coil or arc |
| $L$ | characteristic length |
| $m$ | mass |
| $p$ | pressure |
| $q$ | charge |
| $\tilde{r}$ | coordinate vector |
| $\tilde{r}_c$ | radius of cathode |

| | |
|---|---|
| R | gas constant |
| t | time |
| T | temperature |
| $u_o$ | characteristic velocity |
| $u_{max}$ | maximum velocity along the axis of the arc |
| $\underset{\sim}{v}$ | velocity vector |
| $\varepsilon_o$ | permittivity of free space |
| $\mu$ | molecular viscosity |
| $\mu_o$ | magnetic permeability of free space |
| $\rho$ | density |
| $\sigma$ | electrical conductivity |
| $\tau$ | stress tensor |

COMPUTER SIMULATION OF SOLIDIFICATION

William C. Erickson

Los Alamos Scientific Laboratory

Los Alamos, New Mexico 87545

ABSTRACT

The current state-of-the-art of computer simulation of solidification has been reviewed. Brief descriptions are given for methods of handling the latent heat of fusion and analog, finite element, finite difference, and pseudo-steady state technique used to stimulate solidification are defined. Methods of reducing manpower requirements are discussed as well as general purpose heat transfer codes currently available. The paper concludes with several examples of solidification simulation including die casting, welding, and die design applications.

INTRODUCTION

The use of analytical techniques for calculating solidification heat transfer patterns has essentially paralleled the development of growth of computers. Initially, heat transfer calculations were analytical in nature, but assumptions required to allow the equations to be solved mathematically negated its usefulness in engineering problems such as solidification. With the advent of the analog computer, the interest and application of simulation techniques grew. However, it wasn't until the development of the digital computer that numerical techniques became feasible and the use of computer simulation expanded.

Factors such as increased competition from noncast products, health and safety regulations, and increased energy costs are causing the foundry industry to increase its utilization of scientific principles. This not only includes principles of heat transfer but also areas such as fluid dynamics, kinetics, and nucleations theory.[1] With the changing technological atmosphere in the foundry industry, and with the groundwork laid prior to 1970, computer simulation has reached a point where its application to industrial problems can be considered. In the American Foundrymen's Society honorary Charles Edgar Hoyt Memorial Lecture, Ruddle made the following prediction in 1971: ". . . we shall within two decades see the digital computer take over much of risering practice."[2] In discussing this prediction he stated

> . . . it seems likely that in a short while, all one may need to do is feed into the computer a drawing of the casting plus indication of the thermal properties of the mold material and of the alloy being used, its solidification characteristics and the degree of soundness desired in the casting. The computer would then indicate just what would be the most economic risering and gating system to develop the desired degree of soundness and the location of any residual unsoundness.

Obviously, we are not to this point, but then we are only a third of the way to 1991. This paper will review where we are today with respect to the state-of-the-art of solidification simulation. A brief description of analog, finite difference, finite element, and pseudo-steady state simulation techniques will be given along with a discussion of procedures for handling the latent heat of fusion. It is this factor which sets solidification simulation apart from many heat transfer problems. This will be followed by a discussion of the methodology required for economical use of computer simulation. For the economical use of computers, the manpower costs must be reduced through the use of computer graphics and general purpose software. The paper will conclude with several examples of applied simulation. Sand and ingot casting, die casting, and welding are among the areas reviewed.

Thermal properties play an important role in achieving good simulation results but will not be discussed in detail. The simulation results are only as good as the data input. Unfortunately, good thermal data is not readily available for solids near the melting point or for liquid metals. This is especially true for alloys. A 13-volume compilation of thermal properties edited by Touloukian[3] and the ASM publications on thermophysical properties[4,5] are excellent sources of currently available data. Heat transfer coefficients at interfaces also play an important role in achieving good calculated results. Discussions on the role of interfaces and influencing parameters can be found in the literature.[6,7,8] However, despite these referenced sources, useful data is scarce and thermal properties can be a limiting factor in the accuracy attained in simulation work.

## THE SOLIDIFICATION PROBLEM IN HEAT TRANSFER TERMS

As a casting solidifies, heat transfer occurs in three ways - conduction, convection, and radiation. These are defined by the Fourier, Newton, and Stefan-Boltzmann laws, respectively. Steady state solutions of differential equations derived from these laws can be obtained analytically for a number of uniform property (therefore no phase change) problems. However, casting solidification is a transient problem and as such, partial differential equations relating time, temperature, and location must be solved. Again, the analytical solution of transient problems can only be obtained in a limited number of cases which, because of the phase change involved, do not include solidification.

Three techniques used to obtain solutions to the transient solidification are analog, finite difference and finite element procedures. In addition, a pseudo-steady state approximation technique can be used. In the analog approach, the similarity between corresponding thermal and electrical parameters is used. Both finite element and finite difference are numerical techniques; the first is a functional or integral approach while the second is a differential approach. Finally, the pseudo-steady state process utilizes steady state equations and short time steps to approximate the transient problem.

### Heat of Fusion

Before briefly discussing the techniques used to calculate the heat flow patterns during solidification, a discussion of the latent heat of fusion is warranted. Release of the latent heat of fusion during solidification sets solidification simulation apart from many other heat transfer problems. Differential equations describing heat flow assume that the equations are describing a continuous function. This is not the case when solidification occurs and special procedures must be used to account for the large quantity of heat released during the phase change.

Procedures for handling the heat of fusion fall into two categories. The first class of procedures are the postiterative methods in which the temperatures are adjusted after the calculations are made for that time step. The second class involves accounting for the latent heat of fusion during the calculations for a particular time step.

One of the oldest methods devised to account for the heat of fusion was reported by Eyers et al.[9] This postiterative method consists of transforming the latent heat of fusion into an equivalent number of degrees by dividing the latent heat of fusion by the specific heat. After each time step, the temperatures for nodes that fall below the melting point are reset to the melting point and the difference tabulated for that node. When the required number of degrees have been accumulated for a node, the temperature is allowed to continue dropping.

This procedure is particularly well suited for congruent melting materials, but algorithms can be defined to allow the procedure to work for materials which melt over a temperature range.

A second procedure which is used in the SINDA heat transfer code[10] is based on defining a heat loss rate for the heat of fusion. This rate, defined in energy units/degree, is used to adjust the temperatures after each time step based on the temperature drop. To do this, temperatures at the start of the just completed time step and heat loss rate for each node must be known. This method requires a minimum melting range of one degree which from an engineering standpoint is representative of congruent melting materials.

The heat loss rate method results in a linear release of the latent heat. If this causes problems for alloy systems in which a large volume of the solid forms over a small portion of the solidification temperature range, the latent heat of fusion can be divided and each part released over its own temperature range.

Postiterative procedures do not cause problems from a mathematical standpoint for finite difference or pseudo-steady state procedures. However, researchers at LASL[11] have found that too large a correction in node temperatures causes unacceptable distortions in the heat balances. This was overcome by controlling the time step so that temperature drops were within calculated limits.

The modified specific heat method[12,13] is one of the most popular methods for handling the heat of fusion and is satisfactory for all the basic types of heat transfer calculations described. Here, the latent heat of fusion is converted into the appropriate units for specific heat and added to the specific heat term over the temperature range at which it is to be liberated. This is done such that the increased area under the specific heat vs. temperature curve is equal to the latent heat of fusion. Thus, the latent heat is automatically accounted for during the time step calculations. This method is particularly well suited for alloys which melt over a relatively wide temperature range. It should not be used for congruent melting materials as the probability of the node temperature dropping from above the melting point to below the melting point in one time step is too large. If this occurs, the latent heat of fusion will not be taken into account for that node.

Investigators have used this technique for pure metals[14] by expanding the melting temperature and the error introduced has been acceptable. However, it should be noted that there are significant variations in the thermal arrest portion of the cooling curve.[15]

The manner in which the heat of fusion is added to the specific heat can also play an important role.[16] Figure 1 shows that significant variations occur in the calculated length of time liquid is present in steel ingots and that the differences are a function of the shape of the adjusted specific heat curve.

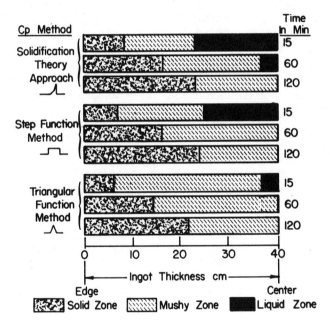

Figure 1: Effect of shape of latent heat function on calculated results.[16]

The discussion to this point has dealt with methods that can be applied independent of the computer program used. There are also many methods which can be incorporated directly into the heat transfer equations and subsequent computer code. One such method was suggested by Eyers et al.[9] and implemented by Sarjant and Slack.[17] This procedure involved defining (1) a modified temperature scale and (2) the specific heat in terms of enthalpy and temperature. By substituting into the basic partial differential equation, the enthalpy, not the temperature, was calculated using numerical procedures. The temperature can then be calculated using enthalpy vs. modified temperature curves. Since latent heat is a portion of the enthalpy of a material, it is automatically accounted for in the calculations. Other special methods can also be found in the literature.[12,18-21]

## Analog Technique

Four methods of solving the heat flow equations were previously mentioned. The first of these, the analog technique, is based on the similarity between the differential equations describing the principles of heat flow, $\frac{\partial T}{\partial t} = \alpha \frac{\partial^2 T}{\partial x^2}$, and the flow of electricity, $\frac{\partial E}{\partial t} = \frac{1}{RC} \frac{\partial^2 E}{\partial x^2}$. In these equations, '$\alpha$' is the thermal diffusivity, 'R' = resistance and 'C' = capacitance. Further comparisons between the two systems are presented in Table I.

TABLE I

COMPARISON OF THERMAL AND ELECTRICAL PROPERTIES[22]

| Thermal Property | Electrical Equivalent |
|---|---|
| Thermal Capacity | Electrical Capacity |
| Thermal Resistivity | Electrical Resistivity |
| Temperature | Voltage |
| Rate of Heat Flow | Current |
| Heat Content | Electric Charge |

Once these similarities are recognized, electrical circuits can be constructed that represent the thermal systems to be simulated and the current measured (the analog of heat flow) at the desired locations. Current (heat) sources/sinks can also be included in the circuit. Ruddle gives a good review of the analog technique for simulating solidification in his book, <u>The Solidification of Castings</u>.[22]

One of the original generalized analog computers built in the United States was the "Heat and Mass Flow Analyzer" at Columbia University.[23] Using this unit, Professor Victor Paschkis performed heat transfer analyses on iron cast in sand in 1944. Under research grants from the American Foundrymen's Society,* Professor Paschkis performed generalized solidification studies[24] and studied such specific problems as temperature drops in pouring ladles,[25] solidification patterns in white irons,[27] and the effect of interface temperature on solidification.[28]

Since the development of the digital computer, analog techniques are used sparingly and in very specialized situations. One such application reported in the literature involved laying out water lines in die casting dies.[29] However, while the principles are the same, this particular application used electrically sensitive paper instead of the analog computer. A general description of this analog technique can be found in the book by Dusinberre.[30]

### Finite Difference Technique

Numerical techniques have long been known, but it was not until the advances made in the digital computer that they became usable. This was because of the time-consuming iterative procedures required to solve the equations.

The numerical analysis equations are formed by expanding the partial differential equations using Taylor's Expansion Theorem. This can result in either center, forward, or backward-differencing equations. Center differencing is preferred because the inherent error is less than for the other two methods. This equation can be solved either explicitly or implicitly. Explicit techniques have the advantage that all temperatures are known at the time the calculations are made. However, a maximum time step exists which if exceeded will cause the equation to represent a thermodynamically impossible condition and the problem becomes numerically unstable. Implicit techniques are unconditionally stable, but they require that a series of simultaneous equations be solved. Dusinberre discusses finite difference techniques for general heat transfer problems and also includes a section on handling latent heat.[30] An excellent discussion of finite difference techniques as applied to casting simulation is given in the AFS monograph Computer Simulation of Solidification.[13]

---

*The American Foundrymen's Society has actively supported computer simulation work since the early 1940's. A review of this and other U. S. work was presented at the Sheffield International Conference on Solidification and Castings.[26]

Problems modeled using finite difference techniques are divided into a series of uniform property, uniform temperature nodes (Figure 2). The node temperature is assumed to be the temperature at the geometric center of the node. In thermal analog terminology, these nodes have a capacitance (ability to store heat) and are connected by conductors which can represent heat transfer by conduction, convection, or radiation.

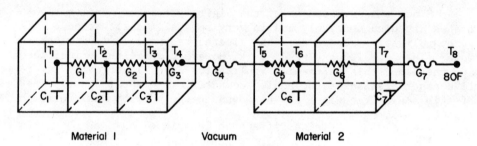

Figure 2: Finite difference modeling approach for thermal analog heat transfer codes.

## Finite Element Technique

Finite element techniques were originally developed by solving complex stress analysis problems. During the 1960's, the applicability of this technique to both steady state[31] and transient[32,33] heat transfer problems was reported. The fact that the technique was originally developed for stress analysis makes finite element solutions particularly appealing to researchers and engineers interested in calculating thermally induced stresses. The fact that the basic elements can be defined independent of thermal properties is also appealing in that the problem can be modeled based on information required without being concerned about relative thermal properties.

A wide variety of element shapes has been demonstrated for stress analysis programs and theoretically all can be used in thermal applications. For a two-dimensional problem using a basic triangular element, the problem is defined as a series of triangular elements with node (temperature) points located at each corner (Figure 3). Thus each node represents a point on all common elements. The volume associated with each node is equal to one third of each triangle. In the simplest form the temperature distribution is assumed to be linear throughout the entire element, but quadratic, cubic, etc., variations could just as well be assumed.

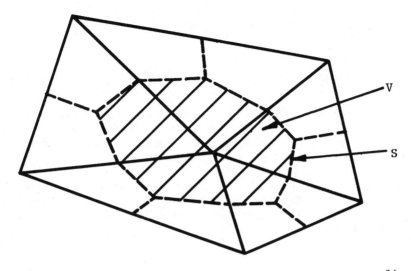

Figure 3: A finite element modeling configuration.[34]

Matrix algebra techniques are used to calculate heat fluxes and the temperature of each node point.[34] Once the variables for the individual elements are determined, matrix techniques are again used to solve and determine the final answer for the "global" or entire matrix. The general pattern required for solution of finite element problems is outlined in the book, The Finite Element Method in Engineering Science, by O. C. Zienkiewicz.[35]

## Pseudo-Steady State Technique

The pseudo-steady state technique is a second finite difference approach to the transient heat conduction problem. However, instead of using partial differential equations to arrive at the answer, steady state heat transfer equations are used to calculate the net heat transfer rate for each node. The heat transfer rate is then converted into the temperature change. In this manner, the individual can work with the simpler algebraic equations and not have to derive numerical solutions from partial differential equations.

The three steady state equations listed below describe heat transfer by convection, conduction, and radiation:

Convection: $\quad q = \bar{h}A(T_1 - T_2)$

Conduction: $$q = \frac{-kA}{\ell}(T_1-T_2)$$

Radiation: $$q = \varepsilon\sigma A(T_1^4-T_2^4)$$

where 'q' = heat flow rate, '$\bar{h}$' = heat transfer coefficient, 'A' = cross sectional area through which heat is transferred, '$\ell$' = distance heat transferred, 'k' = thermal conductivity, '$\varepsilon$' = emissivity, and '$\sigma$' = Stefan-Bolzmann constant.

Kreith[36] shows that the equations for conduction and radiation can be converted to the same form as the convection equation by defining the thermal conductance, 'K,' for the three nodes of heat transfer as:

Convection: $$K = \bar{h}A$$

Conduction: $$K = \frac{-kA}{\ell}$$

Radiation: $$K = \varepsilon\sigma A \left(\frac{T_1^4-T_2^4}{T_1-T_2^1}\right)$$

By using small time steps, heat flow rates between points in transient problems can be approximated using the convection equations and the above thermal conductances. Using the relationship $q = [V\rho(T_f-T_i)C_p]/\Delta t$, where 'V' = node volume, '$\rho$' = density, '$C_p$' = specific heat, '$\Delta t$' = time step, '$T_f$' = final temperature, and '$T_i$' = initial temperature, the new temperature is calculated.

Gatecliff used this approach in simulating the operation of a gas compressor.[37] An integral part of this simulation was calculating heat flow during operation of the compressor. The interesting feature about this work is that a large supercomputer was not required. Gatecliff ran this program on an IBM 1130 with 16,000 words of memory. Subsequently this technique has been applied by Weatherwax and Riegger[38] to die casting simulation. This work will be discussed in greater detail under simulation applications.

The only mathematic or computer code restrictions on how the problem is modeled is related to the time step. Thus, the node size is a function of the information required, the expected temperature gradients in the part, and the time step desired. In the work by Weatherwax and Riegger, the time step was read as input data. Since this is an explicit finite difference approach, a maximum allowable time step exists. This time step must be chosen so that using the

worst case, $\sum_{i=1}^{i=n} K_i (\frac{\Delta t}{V_i \rho C_p}) \leq 1$ where 'K' is the conductance

value in the convection equation and 'n' is the number of neighboring nodes connected to the node in question.

The Biot number, Bi, can give an indication if the node volume is of the proper size. The Biot number is a dimensionless number defined as $Bi = \frac{\overline{h}L}{K_s}$, where '$\overline{h}$' is the average surface conductance, '$k_s$' is the thermal conductivity of the solid, and 'L' is the "significant length" defined as the volume divided by the surface area. If the Biot number is less than 0.1, uniform temperature assumption will introduce errors of less than 5%.

When used for nodes of different materials, Veynik defines a new number K to eliminate a potential conflict in definitions.[39] Both are defined by the same equation, but in the strictest sense of the definition, '$\overline{h}$' designates the heat transfer rate between two materials for the Biot number whereas for K, '$\overline{h}$' also includes the effective heat transfer rate across the gap formed between the casting and mold.

## FLUID FLOW SIMULATION

Up to this point, all the discussions have related to simulating heat transfer problems. However, in many solidification applications, metal flow and heat loss during this flow is of importance. Currently, the standard assumption of instantaneous mold filling is made. The validity of this assumption depends on the system being simulated. For processes where the filling time is a small fraction of the total solidification time for the casting (or a critical part of the casting), this assumption does not introduce significant errors. However, this is not always true in processes where rapid solidification occurs or where a large temperature loss occurs during mold filling. It is the latter areas where fluid flow simulation will play a major role.

Research is being carried out in the field of hydrodynamics modeling. This work has led to the publishing of general purpose codes such as SOLA-SURF.[40] However, these codes are not universal and further development is required before they are applicable to general mold filling problems.

Some fluid flow simulation is currently being performed. This includes work at McGill University concerning fluid flow in ladles[41] and the simulation of injection molding of plastics at Cornell University.[42]

## MODELING A SOLIDIFICATION PROBLEM

Computer simulation of casting solidification has proceeded to the point where it is leaving the research laboratory and entering the manufacturing world. This means that the question to be answered has changed from "Is computer simulation a reliable approach?" to "Is computer simulation economically feasible?" The latter question covers two distinct areas. The first is computer hardware costs. This is the most visible cost and is one everyone is concerned about. A more subtle but much more costly item is manpower. Dr. Barry Boehm studied a series of large computer projects circa 1970 and projected the cost for future Air Force contracts.[43] He predicted that today (1978) manpower costs would be four times greater than computer hardware costs. Even if his estimates were off by a factor of two, manpower requirements still are the major cost and must be addressed in any program.

One means of reducing the manpower is through the use of general purpose software.[44] By using such software, the time required to generate the data needed for simulation can be greatly reduced and a new computer program does not have to be written and debugged for each problem. Manpower can further be reduced by using computer graphics to aid in preparing the data and analyzing the results.

Table II gives a list of typical general purpose heat transfer codes available to the public. ADINAT[45] is a three-dimensional finite element code with working versions in the field, but which has not been totally developed. Its current limitation of constant specific heat and its lack of a procedure for handling latent heat make it unacceptable for solidification studies today. However, it is purported that temperature varying specific heat and a means of accounting for latent heat are being incorporated into a soon-to-be-released version of ADINAT. AYER[34] is a two-dimensional finite element code. This code has the restriction that the latent heat of fusion must be accounted for using the specific heat method or by some other user-defined process. An unnamed finite element code has been written by the University of California, Irving.[48] This code handles two- and three-dimensional problems and isothermal phase changes.

TABLE II

GENERAL PURPOSE HEAT TRANSFER CODES

| Name | | |
|---|---|---|
| ADINAT[45] | Finite Element | Mass. Institute of Technology |
| AYER[34] | Finite Element | Los Alamos Scientific Laboratory |
| --- | Finite Element | Army Cold Regions R&E Lab[50] |
| CINDA-3G,[46] SINDA[10] | Finite Difference | National Aeronautics Space Ad. |
| HEATING5[47] | Finite Difference | Oak Ridge National Laboratory |
| MITAS[48] | Finite Difference | Control Data Corporation |
| TRUMP[49] | Finite Difference | Lawrence Radiation Laboratory |

Perhaps the strongest points in favor of finite element codes for thermal applications are the extensive pre- and post-processors available for these codes. Because of the extensive use of finite element codes for complex structural problems, many of the codes have sophisticated interactive programs for use in modeling the programs. Similarly, post-processing packages with extensive graphics capabilities are also available.

The CINDA family of codes was written for the National Space Aeronautics Administration (NASA). These finite difference codes can handle three-dimensional problems and have extensive program libraries giving the user a wide choice in the type of differencing routine to be used. They handle temperature varying properties and SINDA[10] has a subroutine for handling latent heat. This short subroutine can be used with CINDA-3G.[46] Some graphics output subroutines are available within CINDA-3G and SINDA, but they are not nearly so versatile or sophisticated as those available for finite element codes.

The MITAS[48] code is basically a Martin-Marietta version of CINDA-3G written for Control Data Corporation. The interesting feature of this code is that it is available on Control Data's Cybernet system which makes the code commercially available for users from remote dial-up terminals.

Preprocessors are available for finite difference codes but again they are not so extensive as for finite element codes. Two such meshers are FED[51] written for the TRUMP[49] heat transfer code and HEATMESH-71[52] for the CINDA codes; both are limited to problems with axisymmetric geometries.

Graphics output is a must if large volumes of computer-generated data are to be analyzed accurately and with minimum manpower requirements. This output can take the form of standard cooling curves, isothermal plots, or plots where colors and/or symbols representing temperature ranges are plotted at node locations. When color microfilm is used, the results can be plotted as 35-mm slides or 16-mm movie film.[53]

## APPLICATIONS

Solidification simulation is being used in a wide range of applications. Its biggest use is in the steel industry where multiton castings make experimental programs costly and the gains to be realized from improved processing are large. However, simulation work is not limited to large steel castings. Examples of simulating die casting, welding, centrifugal castings, evaluation of risering systems, and mold development can be found in the literature.

### Sand Ingot Castings

In 1954, Sarjant and Slack published their paper on calculating the internal temperature distribution in steel ingots.[17] In this paper, the authors expanded upon the work of Eyers et al.,[9] and used numerical techniques to calculate transient two-dimensional heat flow patterns. This work formed the impetus for much of the work to follow. A summary of simulation research prior to 1954 is contained in the book by Ruddle.[22]

Both Fischer[54] and Henzel and Keverian[55,56] used General Electric's general purpose transient heat transfer code (THT) to simulate solidification of large steel castings. Excellent agreement was obtained between calculated and experimental data for a 20,000-lb. steel turbine inner-shell casting simulated by Henzel and Keverian.[56] Fischer's work pertained to steel ingots weighing as much as 150 tons in which convection currents played a significant role in the internal heat transfer of the ingots. He accounted for this increased heat transfer rate by multiplying the thermal conductivity of the liquid steel by ten. The fact that a general purpose finite difference computer was used for these works is worthy of mentioning again.

The AFS-sponsored work at the University of Michigan has dealt with the simulation of small ferrous and non-ferrous castings. Once the problem of unknown thermal properties has been overcome, good experimental/calculated agreement has been obtained from data reported in the literature.[13,58,59] Using laboratory-cast aluminum shapes, they have been able to predict the presence of voids and the effect of graphite chills on the solidification pattern.[13,60,61] This work has been extended to a commercial steel casting and the simulation has accurately predicted the presence and location of voids, a means of eliminating the voids, and the shape of the pipe in the hub riser (Fig. 4).[62]

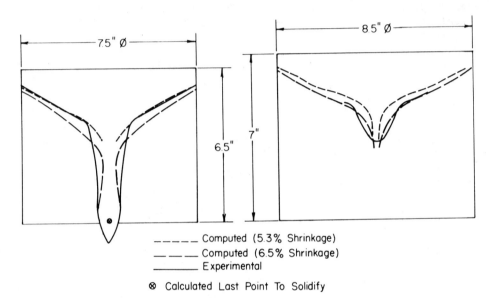

Figure 4: Calculated and experimental pipe shapes in hub-risered steel casting.[62]

Sciama* examined the solidification of aluminum-silicon alloys in permanent molds.[63] He found the enthalpy curve and the interface heat transfer coefficient play important roles in defining the solidification processes. Also, mold wall thickness affects solidification time for thin-walled molds, but does not significantly affect the results as the wall thickness extends beyond a critical

---

*The papers authored by Sciama give a good bibliography of foreign work related to simulation development.

wall thickness. Srinivasan showed similar trends for cast iron cast in metallic molds but also indicated that there was a relationship between casting size and effect of mold wall thickness.[64] The effect of mold wall thickness is also discussed by Roshan et al.[65]

Using the same simulation techniques, Sciama also investigated design parameters for sand-cast eutectic cast iron.[66,67] From these calculations, he was able to show how the thermal center in castings shifted as the thickness ratio of intersecting sections changed. He found that a "fin" can either accelerate or retard the solidification of the attached plate in an L-section. How it behaves depends upon whether the fin thickness is less than or greater than a critical thickness. The influence of core thickness for cast iron, aluminum-silicon, and aluminim-copper alloys was also determined.[68] For cores, the shape of the part, e.g., flat shaped, cylinders, or spheres, determines how the core thickness affects the solidification time.

One means of making computer simulation data more readily available to the working foundryman is to present the computer-generated data in generalized graphical or tabular form. Massey and Sheridan did this for calculated soaking times.[57] Umble uses this technique for cast steel rolls by graphically determining modules reduction factors as a function of roll diameter for varying chill depths.[69] By so doing, these authors were able to make effective use of computer simulation data without the foundry production employees having to work with the computer.

## Die Casting of Aluminum Engine Parts

The die caster is interested in many aspects of the casting cycle. He is concerned about solidification sequence and its effect on casting soundness and temperature gradients in the die are of importance with respect to die life. However, as die temperatures are increased to minimize temperature gradients and extend die life, the heat extraction rate is decreased causing increased cycle time and decreased production rate.

Veynik,[39] Wallace et al.,[70,71] Caswell and Lorentzen,[72] and Thukkaram[73] have dealt with the theoretical aspects of die casting and performed some of the early computer simulation calculations. Work at Penn State University has involved prediction of die failure mechanisms based on calculated thermal distributions for a variety of materials.[74]

Weatherwax and Riegger have published a series of articles on simulating aluminum die casting operations.[38,75,76,77] They calculated the solidification pattern for a 380-aluminum-base alloy connecting rod using production conditions and compared the results

COMPUTER SIMULATION OF SOLIDIFICATION 361

with the macrostructure developed in the casting. The results showed that porosity was present in areas predicted to solidify without proper feeding. Further calculations indicated that die temperature gradients attainable in production would not significantly affect the solidification order, that thermal stock damage is reduced by high initial die temperatures and that directional solidification can be attained with proper die temperature control and gate placement.

Since this initial work, Weatherwax and Riegger have expanded this work to more complex shapes. One such shape is the bore section of an aluminum cylinder block for a small engine, the model of which is shown in Fig. 5.

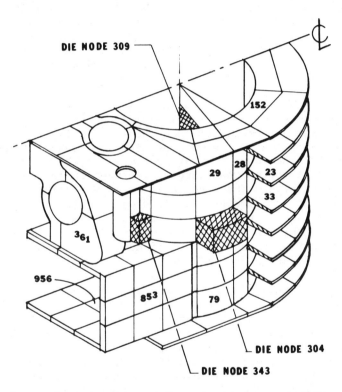

Figure 5: Node network used for simulating solidification of aluminum cylinder block casting.[38]

This work is significant in two important ways: first, the works are true three-dimensional simulations; second, the work was done using the pseudo-steady state method on a PDP-11 minicomputer. The first publication of this series of papers was made in 1971, and the work required only 16K memory. The net result of these calculations were "The casting quality was upgraded by making the changes in gating and casting procedures as suggested by the simulation results."[75]

## Special Casting Processes

Attempts have been made to simulate a variety of other casting processes lumped under the title of "Special Casting Processes" in this report. One such area is centrifugal casting. Lazaridis used numerical techniques to analyze the thermal history of a centrifugal mold for ferritic cast iron pipe.[78] Both start-up and continuous operation were analyzed. From these studies they found that one of the most important, but least controlled, variables with respect to mold life was the thickness of the insulating, mold coating layer. Solidification histories for pipes cast in these molds were presented in a separate paper.[79]

The vast majority of papers involving computer simulation deal with thermal gradients (profiles) in the mold and metal, or to a lesser extent, thermal stresses incurred during the casting process. An exception is the work of Ebisu in Japan.[80] In this work, the author included calculations based on solidification theory to account for macrostructural features in centrifugally cast alloy steel pipe. The importance of mold coating thickness and properties as reported by Lazaridis was confirmed. Ebisu also found that pour temperature was a major factor in producing a uniform, columnar microstructure in the Ni-Cr steel. This was the result of preventing premature solidification caused by longitudinal temperature gradients during mold filling.

Both growth rate 'R' and temperature gradient 'G' were calculated. These values were then used to calculate dendrite length and as an indicator of the type of structure formed. The calculated G/R ratios indicate the dendrites would form tertiary arms. Further, in castings where the simulations indicated low growth rates, the macrostructure was significantly coarser than for castings which the simulation predicted would have a faster growth rate. Thus, the validity of the calculations was verified.

The simulation of continuous casting has been the subject of several papers, many of which were published in the late 1960's. Irving[81] summarized the analytical work related to continuous casting performed prior to 1966. At that time, he chose to use the "integral profile technique" developed by Hills[82] because of the computer requirements for finite difference techniques. About the same time, Mizikar reported simulation work involving one-dimensional numerical analysis of the continuous coating of steel.[83] He did this by dividing the process into three distinct sections: solidification in the mold, solidification in the spray cooling zone, and solidification in the radiant cooling zone. Sufficient confidence in the results was obtained by the author to allow the calculations to be used in designing cooling systems for their continuous casting units.

More recently, simulation of continuous casting has involved the role of stresses in gap formation. Grill et al., have studied this for steel slabs.[84] Interestingly, they applied a three-dimensional finite difference approach for the heat transfer calculations and combined it with a two-dimensional finite element stress analysis technique. Using calculations from this program, they found that hot spots can form near the corners of the steel slab and that these hot spots are associated with gap breakdown, a condition they hypothesize results in high enough tensile strains to cause rupture of the steel shell. The fact that corner hot spots were found in this model stresses the potential dangers involved in assuming symmetry or adibatic surfaces. It is not known if they would have been calculated in a two-dimensional model, but it is certain that they would not have been found in a one-dimensional model.

Mathew and Brody are also studying the continuous casting process from both thermal and mechanical aspects.[85,86] They have applied these calculations to aluminum and magnesium alloys as well as plain carbon steel.

The importance of gap formation and its effect on the heat transfer coefficient not only plays an important role in continuous castings, but also in consumable arc melting. For this process, Eisen and Campagna calculated that the slab thickness itself was the other major controlling variable.[87] They also found that the rate of addition of material changed the pool depth, but did not affect the local solidification time. Arc melting is a difficult process to model and the authors pointed out that because of assumptions made, the model was "semiquantitative."

## Welding

Welding is a special form of the solidification problem in that it involves simulating melting as well as solidification, it has a moving heat source, and in some instances it has metal added to the system as the simulated process continues. Rosenthal discusses the theoretical aspects of heat flow during welding.[88] References in the literature have also been made to a book by Rykalin[89] on the same subject, and a survey paper by Myers et al.[90]

Both finite difference and finite element techniques have been used in weld simulation. The finite element approach has the advantage of being able to calculate thermal-induced stresses during the welding processes. Hibbitt and Marcal[91] investigated the application of finite element techniques to the arc welding processes. This work incorporated the ability to handle the deposition of molten filler metal during welding. These results showed that radiation losses during welding had a significant effect on the peak temperature attained in the weld metal pool.

Using the basic technique developed by Hibbitt and Marcal, Friedman[92,93] also investigated the simulation of gas tungsten arc welding. In Friedman's work, the heat input was applied either as a surface heat flux or to a thin layer at the weld surface; both methods gave the same results. The heat input area was calculated at each time step by the location of the liquids isotherm on the surface. They reported good agreement with experimental results for temperatures in the solid, but could not get good calculated agreement for the shape of the weld fusion zone.

Paley and Hibbert[94] encountered similar problems, but found that by applying the heat of fusion to all nodes in the fusion zone, they could reproduce the fusion zone shape. However, this requires prior knowledge of the weld zone shape and negates much of the advantages of doing simulation studies. The fact that good experimental/calculated agreement was obtained when the heat input was applied to all nodes in the fusion zone suggests that heat transfer conditions within the weld pool were not adequately simulated. This work made good use of computer graphics including three-dimensional plots of temperature in the weld area. They also made isothermal plots which they were then able to correlate with the macrostructure obtained in butt, double-V, three-wire submerged arc, and fillet welds.

Passoja has applied theoretical calculations to electron beam welding[95] and simulation techniques should be applicable to this process. However, because of the high depth-to-width ratio of many electron beam welds, the problem of weld pool heat transfer becomes even more significant. Not only is the means by which the heat is input in the model important, but the extremely high weld metal temperatures require that metal vaporization must also be taken into account in certain welds.

Numerical techniques have also been applied to spot welding.

# COMPUTER SIMULATION OF SOLIDIFICATION

Althey's work on this subject[96] is very theoretical in nature and the correlations presented are between numerical and analytical techniques. However, he does discuss means of handling resistance heating problems.

## Feeding Distance Calculations

The feeding distance of risers is of major concern to metal casters. Feeding distance is not only affected by casting design, but by the manner in which the alloy being cast solidifies. This problem was addressed by Davies[97] in another outstanding example of combining solidification theory with numerical simulation. Using the numerical procedures outlined by Sarjant and Slack, Davies calculated the solidus velocity ($V_s$) along the axis of the plate and solidification curves, i.e., time vs. location plots, for isotherms representing various fractions solid.

The equation for capillary feeding distance, '$l_f$,' was derived and given as $l_f = (Bh\rho r^2)/(m\eta V_s)$ where 'B' is a constant = .22, 'h' = feeding height, '$\rho$' = density liquid metal, 'r' - capillary (pore) radius, 'm' = solidification contraction, '$\eta$' = dynamic viscosity, and '$V_s$' = solidus velocity. Based on the mode of freezing, estimates were made for the volume metal solidified at which feeding problems would occur. For example, Davies estimated that feeding problems would start at 95% solid for 0.6% C steel, 95% solid for 99.99% Al, and 90% solid for 99.6 Al. The capillary length was determined and plotted on the same graph as the calculated feeding distances for each node (Fig. 6). Whenever the feeding distance value is less than the capillary length, centerline shrinkage can be expected.

Davies calculated feeding distances for a wide range of materials and compared them with empirically determined values found in the literature and used in industry. Excellent agreement was obtained. As Davies points out, it is important that reasonable assumptions be made for when feeding problems start, for the results depend on these assumptions.

## Mold Design Evaluation

It was previously stated that there are three means by which manpower requirements could be reduced in computer-related work. An example of this approach is given in solidification simulation work being performed at the Los Alamos Scientific Laboratory.[98] The program's objective is to optimize the design of a coated graphite mold that would significantly reduce the material required to cast hemispherical enriched uranium parts. To accomplish the goal, it is proposed that a recirculating tin pump be used to achieve directional solidification in the cast parts.

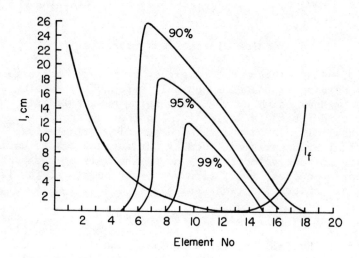

Fig. 6: Calculated feeding distances for aluminum plate.[97]

A modified version of the FED mesher was used to generate node and conductor input data for the CINDA-3G heat transfer code. The calculated results were then plotted for analysis on 16-mm film using a general plotting program.[99] Using this technique, a mold design was developed which promoted the desired directional solidification and reduced the metal requirements by 55%. The work performed for this program is summarized in a 16-mm, computer generated, color film.

## SUMMARY

The technical feasibility of computer simulation of solidification problems has been demonstrated. The calculations can be made using finite element or finite difference techniques which in general, require larger, faster computers. It has also been demonstrated that using steady state approximation techniques, solidification problems can be successfully modeled on smaller computers such as PDP-11. Release of the latent heat of fusion requires special attention, but it can be handled in one of several methods. The method best suited for the problem is a function of how the metal being simulated solidifies and the computer code being used.

A wide variety of metal systems and types of solidification processes have been simulated. Its largest use has been in the casting of large steel shapes. However, it has also been demonstrated that computer simulation can be used for smaller castings to detect defects and to determine the potential benefit of suggested engineering changes. Computer simulation work has been applied to both slow and fast cooling systems. The former is represented by sand castings, and the latter by die casting and welding.

The works by Ebisu and Davies demonstrate how computer simulation can be combined with solidification theory and scientific principles to study casting problems. Ebisu predicted the macrostructure developed in centrifugal castings while Davies calculated feeding distances for risers, taking into account the manner in which the metal solidified.

Simulation of fluid flow during casting processes is still in its infancy. Work has begun in simulating metal flow and it is anticipated that this will be the next area of major technological advances in the simulation field.

General purpose software must be used if solidification simulation is to be economically employed in the foundry industry. It is this area which must be developed if Ruddle's prediction quoted in the Introduction is to materialize by 1991. The general techniques needed to produce this software are available, but large expenditures of money will be required to implement them. It is questionable if the preprocessor (mesher) computer programs will be developed specifically for simulation problems because of economic reasons. It is more likely that the thermal simulation software will be incorporated into modified CAD/CAM systems. In this way, computer-aided design programs could be used for inputting and manipulating the basic geometric data and software would only have to be written for performing the specific functions required for thermal and fluid flow simulation.

With the rapid advances made in the areas of CAD/CAM programs, the 1991 date of Ruddle is entirely realistic. In fact, if the work continues at the present pace, his goal will be reached earlier than predicted.

## ACKNOWLEDGEMENTS

Several individuals contributed to the preparation of this paper. From the Los Alamos Scientific Laboratory, the help of

J. M. Dickinson, R. E. Keenan, G. A. Bennett, and the staff of the LASL Technical Library, in particular M. H. Treiman, was greatly appreciated. Fellow members of the AFS Heat Transfer Committee, J. T. Berry, R. D. Ruddle, O. K. Riegger, A. E. Umble, and L. J. D. Sully, also contributed through discussions and aid in locating references not available to the author.

## References

1. G. S. Cole, "Solidification Science and Foundry Technology," AFS International Cast Metals Journal, Vol. 2, 33-38 (June 1977).

2. R. W. Ruddle, "Risering - Past, Present and Future," Transactions, American Foundrymen's Society, Vol. 79, 269-280 (1971).

3. Thermophysical Properties of Matter, Y. S. Touloukian, Ed., (IFI/Plenum, New York, 1970).

4. Selected Values of the Thermodynamic Properties of the Elements, prep. by R. Hultgren, P. D. Desai, D. T. Hawkin, M. Gleiser, K. K. Kelley, and D. E. Wagman, American Society for Metals, Metals Park, Ohio (1973).

5. Selected Values of the Thermodynamic Properties of Binary Alloys, prep. by R. Hultgren, P. D. Desai, D. T. Hawkins, M. Gleiser, and K. K. Kelley, American Society for Metals, Metals Park, Ohio (1973).

6. R. D. Pehlke, "The Interface in Computer Simulation of Heat Transfer in Metallurgical Processes," Metals Engineering Quarterly, Vol. 11, No. 3, 9-13 (August 1971).

7. L. J. D. Sully, "The Thermal Interface Between Castings and Chill Molds," Transactions, American Foundrymen's Society, Vol. 84, 735-744 (1976).

8. D. R. Durham and J. T. Berry, "The Role of the Mold-Metal Interface During Solidification of a Pure Metal Against a Chill," Transactions, American Foundrymen's Society, Vol. 82, 101-110 (1974).

9. N. R. Eyers, D. R. Hartree, J. Ingham, R. Jackson, R. J. Sarjant, and J. R. Wagstaff, "The calculations of Variable Heat Flow in Solids," Philosophical Transactions of the Royal Society of London, Series A, Vol. 240, 1-57 (1948).

10. J. D. Gaski, L. C. Fink, and T. Ishimoto, "Systems Improved Numerical Differencing Analyzer Users' Manual," TRW Systems, Redondo Beach, California (1970).

11. G. A. Bennett, Los Alamos Scientific Laboratory, personal communication, May 1978.

12. G. Comini, S. DelGuidice, R. W. Lewis, and O. C. Zienkiewicz, "Finite Element Solution of Non-Linear Heat Conduction Problems with Special References to Phase Change," International Journal of Numerical Methods in Engineering, Vo. 8, 613-624 (1974).

13. R. D, Pehlke, R. E. Marrone, and J. O. Wilkes, Computer Simulation of Solidification, (American Foundrymen's Society, DesPlaines, Illinois, 1976).

14. R. D. Pehlke, M. J. Kirt, R. E. Marrone, and D. J. Cook, "Numerical Simulation of Casting Solidification," AFS Cast Metals Research Journal, Vol. 9, No. 2, 49-55 (June 1973).

15. W. C. Erickson and A. V. Houghton, "Simulating the Effect of Thermocouple Assemblies on Temperature Measurements, American Foundrymen's Society, Vol. 85, 56-64 (1977).

16. A. I. Koler, J. D. Thomas, and A. A. Tzavaras, "Computations of Heat of Fusion in a Mathematical Model for Large Steel Cast Shapes," AFS Cast Metals Research Journal, Vol. 9, No. 4, 156-159 (December 1973).

17. R. J. Sarjant and M. R. Slack, "Internal Temperature Distribution in the Cooling and Reheating of Steel Ingots," Journal of the Iron and Steel Institute, Vol. 177, 428-444 (August 1954).

18. A. Lazaridis, "A Numerical Solution of the Multidimensional Solidification (or Melting) Problem," Int'l Journal of Heat and Mass Transfer, Vol. 13, 1459-1477 (1970).

19. E. Friedman, "An Iterative Procedure for Including Phase Change in Transient Heat Conduction Programs and Its Incorporation Into the Finite Element Method," Nat'l Heat Trans. Conference, Salt Lake City, Utah (August 1977).

20. O. C. Zienkiewicz, C. J. Parakh, and A. J. Wills, "The Application of Finite Elements to Heat Conduction Problems Involving Latent Heat," Rock Mechanics, Vol. 5, 65-76 (1973).

21. C. Bonacina, G. Comini, A. Fasano, and M. Primicerio, "Numerical Solution of Phase-Change Problems," Int. J. Heat and Mass Transfer, Vol. 16, 1825-1832 (1973).

22. R. W. Ruddle, The Solidification of Castings, 2nd Edition (The Institute of Metals, London, 1957).

23. V. Paschkis, "Heat Flow Problems in Foundry Work," Transactions, American Foundrymen's Association, Vol. 52, 649-670 (1944-45).

24. V. Paschkis and J. W. Hlinka, "Some Generalized Solidification Studies," Transactions, American Foundrymen's Society, Vol. 65, 222-227 (1957).

25. V. Paschkis and J. W. Hlinka, "Temperature Drop in Pouring Ladles, Part II," Transactions, American Foundrymen's Society, Vol. 65, 276-280 (1957).

26. R. D. Pehlke, J. T. Berry, W. C. Erickson, and C. H. Jacobs, "The Simulation of Shaped Casting Solidification," Int. Conf. Solidification and Casting (Sheffield, England, July 1977).

27. V. Paschkis, "Studies of Solidification of White Iron Castings," Transaction, American Foundrymen's Society, Vol. 56, 371-373, (1948).

28. V. Paschkis and J. W. Hlinka, "Some Remarks on the Relationship of Interface Temperature and Solidification," Transactions, American Foundrymen's Society, Vol. 66, 213-221 (1958).

29. G. K. Ruhlandt, "Thermal Analysis of Die Casting Dies," Modern Castings, p. 75 (August 1967).

30. G. M. Dusinberre, Heat Transfer Calculations by Finite Differences, (International Textbook Company, Scranton, PA, 1961).

31. O. C. Zienkiewicz and Y. K. Cheung, "Finite Elements in the Solution of Field Problems," The Engineer, Vol. 220, 507-510 (1965).

32. O. C. Zienkiewicz and C. J. Perekh, "Transient Field Problems: Two-Dimensional and Three-Dimensional Analysis by Isoparametric Finite Elements," Internal Journal for Numerical Methods in Engineering, Vol. 2, 61-72 (1970).

33. E. L. Wilson and R. E. Nickell, "Application of the Finite Element Method to Heat Conduction Analysis," Nuclear Engineering and Design, Vol. 4, 276-286 (1966).

34. R. G. Lawton, "The AYER Heat Conduction Computer Program," Los Alamos Scientific Laboratory report LA-5613-MS (April 1974).

35. O. C. Zienkiewicz, The Finite Element Method in Engineering Science (McGraw Hill, London, 1971).

36. F. Krieth, Principle of Heat Transfer, 3rd Edition (Intext Educational Publications, New York, 1973).

37. G. W. Gatecliff, "A Digital Simulation of a Reciprocating Hermetic Compressor Including Comparisons with Experiment," Ph.D. thesis, University of Michigan (1969).

38. R. Weatherwax and O. Riegger, "Solidification Studies of Die Cast Aluminum Small Engine Crankcase Casting," Transactions, Society of Die Casting Engineers, Paper No. G-T77-016 (1977).

39. A. I. Veynik, Theory of Special Casting Methods (Translated by American Society of Mechanical Engineers, New York, 1962).

40. C. W. Hirt, B. D. Nichols, and N. C. Romero, "SOLA-A Numerical Solution Algorithm for Transient Fluid Flows," Los Alamos Scientific Laboratory report LA-5852 (1975), LA-5852 Addendum (1976).

41. M. Salcudean and R. I. L. Guthrie, "Fluid Flow in Filling Ladles," to be published, Metallurgical Transactions B (1978).

42. K. S. Wang, S. F. Shen, J. F. Stevenson, and C. A. Hieber, "Computer-Aided Injection Molding System, Progress Report No. 4," Cornell University (September 1977).

43. B. W. Boehm, "Software and Its Impact: A Quantitative Assessment," Datamation, 48-59 (May 1973).

44. W. C. Erickson, "The Use of General Purpose Software in Casting Solidification Simulation," Metallurgical Transactions B, Vol. 88, 93-97 (March 1977).

45. K. Bathe, "ADINAT, A Finite Element Program for Automatic Dynamic Incremental Nonlinear Analysis of Temperature," Massachusetts Institute of Technology report 82448-5 (May 1977).

46. D. R. Lewis, J. D. Gaski, and L. R. Thompson, "Chrysler Improved Numerical Differencing Analyzer for 3rd Generation Computers," Chrysler Corporation Space Division, New Orleans, TN-AP-67-287 (1967).

47. W. D. Turner, D. C. Elrod, and I. I. Siman-Tov, HEATING5: An IBM 360 Heat Conduction Program," Union Carbide Corporation, Nuclear Division, report No. ORNL/CSD/TM-15, Oak Ridge, Tennessee (March 1977).

48. <u>MITAS User Information Manual</u>, Pub. #86615000, Control Data Corporation, Minneapolis, Minnesota (1973).

49. A. L. Edwards, "TRUMP: A Computer Program for Transient and Steady State Temperature Distribution in Multidimensional Systems," Lawrence Radiation Laboratory report UCRL-14754, Rev. 2 (1969).

50. G. L. Guymon and T. V. Hromadka II, "Finite Element Model of Transient Heat Conduction with Hisothermal Phase Change (Two and Three Dimensional)," Cold Regions Research and Engineering Laboratory report No. CRREL-SR-77038, Hanover, New Hampshire (November 1977).

51. D. A. Schauer, "FED: A Computer Program to Generate Geometric Input for the Heat-Transfer Code TRUMP," Lawrence Livermore Laboratory report UCRL-50816-Rev. 1 (January 1973).

52. V. K. Gabrielson, "HEATMESH-71: A Computer Code for Generating Geometrical Data Required for Studies of Heat Transfer in Axisymmetric Structures," Sandia Laboratories report SCL-DR-720004 (September 1972).

53. W. C. Erickson, "The Use of Computer-Generated Color Movies in Simulating Casting Solidification," Transactions, American Foundrymen's Society, Vol. 82, 161-164 (1974).

54. G. A. Fischer, "Solidification and Soundness Predictions for Large Steel Ingots," Proceedings of the American Society for Testing and Materials, Vol. 62, 1137-1153 (1962).

55. J. G. Henzel, Jr., and J. Keverian, "Predicting Solidification Patterns in a Steel Valve Casting by Means of a Digital Computer," Metals Engineering Quarterly, Vol. 5, No. 2, 39-44 (May 1965).

56. J. G. Henzel, Jr., and J. Keverian, "The Theory and Application of a Digital Computer in Predicting Solidification Patterns," Journal of Metals, Vol. 17, No. 5, 561-568 (May 1965).

57. I. D. Massey and A. T. Sheridan, "Theoretical Predictions of Earliest Rolling Times and Solidification Times of Ingots," Journal of the Iron and Steel Institute, Vol. 209, 391-395 (May 1971).

58. R. E. Marrone, J. O. Wilkes, and R. D. Pehlke, "Numerical Simulation of Solidification, Part I: Low Carbon Steel Casting - 'T' Shape," AFS Cast Metals Research Journal, Vol. 6, No. 4, 184-188 (December 1970).

59. R. E. Marrone, J. O. Wilkes, and R. D. Pehlke, "Numerical Simulation of Solidification, Part II: Low Carbon Steel Casting - 'L' Shape," AFS Cast Metals Research Journal, Vol. 6, No. 4, 188-192 (December 1970).

60. A. Jeyarajan and R. D. Pehlke, "Casting Design by Computer," Transactions, American Foundrymen's Society, Vol. 83, 405-412 (1975).

61. A. Jeyarajan and R. D. Pehlke, "Computer Simulation of Solidification of a Casting with a Chill," Transactions, American Foundrymen's Society, Vol. 84, 647-652 (1976).

62. R. D. Pehlke and A. Jeyarajan, "Application of Computer-Aided Design to a Commerical Steel Casting," 82nd AFS Casting Congress (Detroit, Michigan, April 1978).

63. G. Sciama, "Computation of Cooling Time in Permanent Mold Cast Cylindrical Bars - Checking Tests," AFS Cast Metals Research Journal, Vol. 4, No. 2, 62-68 (June 1968).

64. M. N. Srinivasan, "Analytical Studies on the Solidification of Hypereutectic Cast Iron in Metallic Molds," AFS Cast Metals Research Journal, Vol. 11, No. 3, 91-99 (September 1975).

65. H. Md. Roshan, E. G. Ramachandran, M. R. Seshadri, and A. Ramachandran, "Analytical Solution to the Heat Transfer in Mold Walls During Solidification of Metals," AFS Cast Metals Research Journal, Vol. 10, No. 1, 39-47 (March 1974).

66. G. Sciama, "Study of the Solidification of Elbows in Cast Iron," AFS Cast Metals Research Journal, Vol. 8, No. 1, 20-24 (March 1972).

67. G. Sciama, "Solidification of Simple Profiles with Molten Hot Spots - L, T, and Cruciform Junctions," AFS Cast Metals Research Journal, Vol. 8, No. 4, 145-149 (December 1972).

68. G. Sciama, "Study of the Solidification of Hollow Castings - Influence of Relative Thickness of Core," Transactions, American Foundrymen's Society, Vol. 80, 141-148 (1972).

69. A. E. Umble, "Computer Simulation for Determining Chilling Practice on Cast Steel Bolts," Transactions, American Foundrymen's Society, Vol. 86, 69-70 (1978).

70. W. F. Stuhrkel and J. F. Wallace, "Gating of Die Castings," Modern Castings, Vol. 44, 51-79 (January 1966).

71. D. Linsey and J. F. Wallace, "Heat and Fluid Flow in the Die Casting Process," Transactions, Society of Die Casting Engineers, Paper No. 12 (1968).

72. B. F. Caswell and P. Lorentzen, "Some Theoretical Aspects of Heat Transfer in the Die Casting Process," Die Casting Engineers, 26-32 (March-April 1968).

73. P. Thukkaram, "Heat Transfer in Die Casting Dies," Transactions, Society of Die Casting Engineers, Paper No. 61 (1970).

74. J. M. Samuals, F. W. Schmidt, and A. B. Draper, "Temperature Distribution and Thermal Stresses in Die Materials Producing Stainless Steel Castings," Transactions, American Foundrymen's Society, Vol. 82, 285-294 (1974).

75. R. B. Weatherwax and O. K. Riegger, "Theoretical and Experimental Studies of Die Casting Techniques for Small Engine Connecting Rods," Transactions, Society of Automotive Engineers, 2112-2119 (1971).

76. R. B. Weatherwax and O. K. Riegger, "Computer Simulation of the Die Casting Machine Cycle," Transactions, Society of Die Casting Engineers, Paper No. G-T77-015 (1977).

77. R. B. Weatherwax and O. K. Riegger, "Computer-Aided Solidification Study of a Die Cast Aluminum Piston," Transactions, American Foundrymen's Society, Vol. 85, 317-322 (1977).

78. A. Lazardis, "Thermal Analysis of Centrifugal Casting Molds," AFS Cast Metals Research Journal, Vol. 6, No. 4, 153-160 (December 1970).

79. A. Lazardis, "Effect of Machine Variables on the Centrifugal Casting of Long Pipes," AFS Cast Metals Research Journal, Vol. 8, No. 2, 77-82 (June 1972).

80. Y. Ebisu, "Computer Simulations on the Macrostructures in Centrifugal Castings," Transactions, American Foundrymen's Society, Vol. 85, 643-653 (1977).

81. W. R. Irving, "Heat Transfer in Continuous Casting Molds," Journal of the Iron and Steel Institute, Vol. 205, 271-277 (March 1967).

82. A. W. D. Hills, "Integral Profile Method for Analysis of Heat Flow on Continuous Casting of Metals," Inst. of Chemical Engineers Symposium on Chemical Engineering on the Metallurgical Industries, 128-149 (1963).

83. E. A. Mizikar, "Mathematical Heat Transfer Model for Solidification of Continuously Cast Steel Slabs," Transactions, Metallurgical Society of AIME, Vol. 239, 1747-1753 (November 1967).

84. A. Grill, K. Sorimachi, and J. K. Brimacombe, "Heat Flow, Gap Formation and Break-outs in the Continuous Casting of Steel Slabs," Metallurgical Transactions B, Vol. 7B, 177-189 (June 1976).

85. J. Mathew and H. D. Brody, "Analysis of Heat Transfer in Continuous Casting Using a Finite Element Method," Conference on Computer Simulation for Materials Applications (Gaithersburg, Maryland, April 1976).

86. J. Mathew and H. D. Brody, "Simulation of Thermal Stresses in Continuous Casting Using a Finite Element Method," Conference on Computer Simulation for Materials Applications (Gaithersburg, Maryland, April 1976).

87. W. B. Eisen and A. Campagna, "Computer Simulation of Consumable Melted Slabs," Metallurgical Transactions, Vol. 1, 849-856 April 1970).

88. D. Rosenthal, "Mathematical Theory of Heat Distribution During Welding and Cutting," The Welding Journal Research Supplement, Vol. 20, No. 5, 221S-234S (May 1941).

89. N. N. Rykalin, The Calculation of Thermal Processes in Welding, (Mashgiz, 1951; translated by Z. Paley and C. M. Adams, Jr., 1963).

90. P. S. Myers, O. A. Uyehara, and G. L. Borman, "Fundamentals of Heat Flow in Welding," Welding Research Council Bulletin No. 123 (July 1967).

91. H. D. Hibbitt and P. V. Marcal, "A Numerical, Thermo-Mechanical Model for the Welding and Subsequent Loading of a Fabricated Structure," Computers and Structures, Vol. 3, 1145-1174 (1973).

92. E. Friedman, "Thermomechanical Analysis of the Welding Process Using the Finite Element Method," Transactions, ASME Journal of Pressure Vessel Technology, Vol. 97, Series J, No. 3, 206-213 (August 1975).

93. E. Friedman and S. S. Glickstein, "An Investigation of the Thermal Response of Stationary Gas Tungsten Arc Welds," Welding Journal Research Supplement, Vol. 55, No. 12, 408S-420S (December 1976).

94. Z. Paley and P. D. Hibbert, "Computation of Temperatures in Actual Weld Designs," Welding Journal Research Supplement, Vol. 54, No. 11, 385S-392S (November 1975).

95. D. E. Passoja, "Heat Flow in Electron Beam Welds," Welding Journal Research Supplement, Vol. 45, No. 8, 379S-384S (August 1966).

96. D. R. Atthey, "A Finite Difference Scheme for Melting Problems," Journal Institute Mathematics and Its Applications, Vol. 13, 353-356 (1974).

97. V. de L. Davies, "Feeding Range Determination by Numerically Computed Heat Distribution," AFS Cast Metals Research Journal, Vol. 11, No. 2, 33-44 (June 1975).

98. R. E. Keenan and W. C. Erickson, "Use of Computers in Mold Design," Proceedings of Army Science Conference (West Point, New York, June 1978).

99. W. C. Erickson, "The Use of General-Purpose Heat-Transfer Code for Casting Simulation," MS Thesis, University of New Mexico (Albuquerque, NM 1975).

# INDEX

Aerospace industry, laser processing in, 72
Abrasives
  alloy, *see* Alloy abrasives
  characterization features of, 219–223
  commercial forms of, 218–219
  cubic boron nitride as, 244–248
  differential growth rates of, 221
  manufacture of, 218–219
  microhardness of, 222
  nonequilibrium conditions for, 220
  phase diagram for, 219
  quantitative microstructure of, 222
  relative performance trends in, 217
  for steel conditioning, 216
ADINAT finite element code, 356
Alloy abrasives
  cracks in, 226
  manufacture of, 218–219
  microfracture of, 223
  microhardness of, 222
  quantitative microstructure of, 222
Alloys, homogeneity in, 278
Alumina
  co-fusion with zirconia, 215
  cubic polymorph of, 221
  dendritic "cubic" form of, 220
  as grinding material, 215
Alumina-zirconia abrasives, 215–217
Alumina-zirconia alloys
  composition of, 218
  differential growth rates in, 221
  heat flow through, 221
  phase diagram for, 219
Aluminum alloys
  butt welding of, 122
  laser-shocked, 67
  laser weld characteristics in, 122
  permeability of, 311
Aluminum cylinder block casting, solidification simulation in, 361
Aluminum engine parts, solidification simulation for, 360–362
Aluminum-silicon alloys, solidification in, 359
Aluminum substrate, liquid-solid interfaces during melting of, 39
American Foundrymen's Society, 351
Amphère's law, 320–321
Analog technique, for heat flow equations, 350–351
Annealing, intercritical, 175, 191–197
Arc, free-burning, 337, *see also* Electric arc
Arc welding, plasma dynamics in, 341
Atomization
  heat flow during, 14–22

Atomization (continued)
  two-fluid, 101
Atomization mechanisms, in power processing, 101–106
Atomization methods
  centrifugal, 97–99
  gas, 92–94
  in powder processing, 92–100
  roller, 99–100
  steam, 96–97
  ultrasonic, 99
  vacuum or soluble gas, 99–100
  water, 94–97
Ausaging, temperature increase in, 200–203
Austenite
  mechanical transformation of, 190
  retransformed, 192
  reversion to martensite, 187, 197–209
  thermal transformation of, 190
  transformation strengthening of, 199–204
Austenite precipitation
  carbide dissolution in, 182
  during intercritical tempering, 187
Austenite reversion, 187, 197–209
Austenite steel, see also Steel(s)
  carbon as stabilizing element in, 179–181
  transformation strengthening of, 176
Autogenous laser butt welds, fatigue properties of, 121
Automotive industry, laser processing in, 71–72
Average cooling rate, 19, 24
AYER finite element code, 356

B-implanted silicon, laser- vs. thermally annealed, 65
Biot numbers, 15, 18, 355
  average cooling rate and, 24
  solid-liquid interface velocity and, 19, 21

Biot numbers (continued)
  temperature gradients and, 24
Boltzmann equation, 335
Boron, laser annealing and, 65–66
Boron-implemented silicon, 64–65
Boron nitride, cubic, 244–248
Bulk rapid solidification, laser beams in, 61–64
Butt welding, of aluminum alloys, 122

Cabot Corporation, 278
Carbon, as austenite-stabilizing element in steel, 179–181
CBN, see Cubic boron nitride
Centrifugal atomization, in powder processing, 97–99
Centrifugal casting, solidification insulation in, 362
Chemical vapor deposition, laser-controlled, 70
Chip removal process, basic characteristics of, 259
CINDA heat transfer code, 357, 366
Circular region, temperature at center of, 41
Coble creep, 156
Coil current/velocity relationships, in induction furnace, 323–325
$CO_2$ laser welds, see also Laser welding
  fusion zone purification in, 124–125
  steel penetration in, 113
Columbia University, 351
Computer simulation, of solidification, 345–367
Conduction/induction current ratio, 321
Continuous cooling, simulation of, 362–363

# INDEX

Cooling rate
  average, 19
  heat flux and, 35-36
Creep
  Coble, 156
  diffusion-controlled dislocation, 157
  slip, 157, 163
  stress-directed diffusional, 156
Creep feed grinders, 242
Creep feed grinding, 237-241
  surface temperatures in, 240
Creep flow, superplasticity and, 147
Cryogenic steels
  intercritical annealing of, 191
  2B cycling treatment for, 210-212
"Cryonic 5" steel, thermal treatment of, 210
Crystalline solidification
  melt depth in, 32
  process variables/time factor in, 19
  undercooling prior to, 18
Cubic boron nitride
  as abrasive material, 244-248
  compared with diamond, 245
  grinding wheel of, 246-247
Cylindrical remelted ingots, macrosegregation and interdendritic flow in, 284-287

Darcy's law, 281, 298-299
DBTT, see Ductile-brittle transition temperature
Deep penetration laser welding, 49, 113, see also Laser welding
  filler material in, 124
  schematic of, 50
Dendrite arm spacing
  cooling rate and, 106
  heat transfer calculations for, 17
Diamond, as abrasive, 244

Die casting, solidification simulation in, 360
Dimensionless cooling rate, and melting point of noncrystalline solidification of aluminum melt, 27
Dimensionless instantaneous average cooling rate, 18
Dimensionless interface velocity, for Newtonian solidification, 20
Dimensionless temperature distributions, in liquid droplets, 16
Droplet generation, electrohydrodynamic method for, 80
Dual-phase steels
  development of, 191-192
  microstructure of after intercritical annealing, 195
  post-necking elongation in, 197
  stress-strain curve for, 197-198
  superiority of, 194
  ultimate tensile strength of, 193
Ductile-brittle transition temperature, 177
  lowering of, 179
Durarc® process, 97

EHD processes, see Electrohydrodynamic processes
Electrical circuits, for thermal system representation, 350
Electric arc
  in metals processing, 334-341
  principles of, 338-339
Electric arc melting, plasma dynamic principles in, 339-340
Electrohydrodynamic apparatus
  experimental results with, 82-87

Electrohydrodynamic apparatus (continued)
  microdroplet generation source in, 81
  splat cooling studies with, 85–87
Electrohydrodynamic metal processing, 79–88
  for droplet generation, 80
  for tabletop rapid quenching metal powder generation, 79
Electromagnetic force field
  calculation of, 320
  in ESR system, 312–314
Electron beam welding, simulation techniques in, 364
Electroslag remelted ingots
  electromagnetic force field in, 312–314
  freckles in, 295–298
  interdendritic liquid density in, 279
  macrosegregation in, 273–313
  metal pool profiles in, 291
  "solidification induced" convection in, 280
Electroslag welding, 329–334
  principle of, 329–330
  schematic diagram of, 331
  streamline pattern for, 332
  surface velocities in, 330
  velocity vector for, 333
End milling, working region for, 273
ESR ingots, see Electroslag remelted ingots
ESW, see Electroslag welding
Experimental laser processing, operational regimes in, 53

Faraday's law, 320
Feeding distance calculations, in solidification simulation, 365
Ferrous alloys, laser weld characteristics in, 117–120

Filler material, in laser welding, 123–124
Fine structure superplastic flow, 149–161, see also Superplasticity
  creep behavior in, 160
  grain size in, 159
  phenomenological equation for, 156–157
  prerequisites for, 152–155
Fine structure superplastic materials, strain rate and, 165
Finite diffuse technique, for heat flow equations, 352
Finite element codes, 356
Finite element modeling configuration, 353
Finite element technique, for heat flow equations, 352–353
Fluid flow simulation, solidification applications and, 355–356
Fracture, strain rate sensitivity experiment and, 141
Fracture surface, austenite distribution and, 188–189
"Freckles"
  in ESR ingots, 277
  formation of, 290, 294–304
Fusion, heat of, 348
Fusion zone purification, in laser welding, 119, 124–127

Gas atomization
  mechanism of, 101–103
  operating conditions in, 94
  in powder processing, 92–94
  two-jet configuration in, 93
  variables in, 94
Gas pressure, particle size and, 95
Gaussian heat flux distribution, 28, 40

# INDEX

Grinding, *see also* Abrasives; Material removal processes
  energy conserved in, 235
  exponential coefficient in, 234
  external plunge, 235
  as finishing process, 229
  force generated during, 233
  future of, 230
  high-speed creep feed, 237-238
  micro-chip formation process in, 235
  pendulum operations in, 240-241
  recent advances in, 229-250
  selected examples in, 233
  speed stroke, 241-244
  surface temperature in, 236-237, 240
  wheel speeds in, 237
Grinding force
  analytical description of, 233-235
  depth of cut and, 239
Grinding materials, innovations in, 215-226
Grinding technology, development of, 230-233
Grinding tests, 224-225
Grinding wheel
  cubic boron nitride, 246-247
  for steel conditioning, 216
Grinding wheel hardness
  electro-sonic determination of, 248-250
  Grindo-Sonic test instrument for, 249-250

Heat and Mass Flow Analyzer, 351
Heat flow
  during atomization, 14-22
  during melting and solidification of surface layer, 28-42
  one-dimensional, 30-32, 37
  square root of time in, 30-32
  two-dimensional, 38
Heat flow equations
  analog technique for, 350-351

Heat flow equations (continued)
  finite difference technique for, 351-352
  finite element technique in, 352-353
  pseudo-steady state technique for, 353-355
Heat flow limitations, in rapid solidification processing, 13-42
Heat flow model, adsorbed heat flux and, 28
Heat flux
  cooling rate and, 35-36
  effect of change in, 37
  instantaneous average cooling rate and, 35
  melt depth and, 33-34
  and temperature at center of circular region, 41
  uniform versus Gaussian, 29, 40
Heat flux distribution, position and time as functions of, 28
HEATINGS heat transfer code, 357
Heat loss, fluid flow simulation and 355-356
Heat of fusion
  solidification in, 347-350
  specific heat method for, 348
Heat transfer, solidification and, 347-355
Heat transfer codes, 356-358
Heat transfer coefficients
  of atomization processes, 14
  calculations from DAS, 17
  limitation on achievement of, 15
  for solidification of aluminum against metal substrate, 23
  upper limits for, 22
Heat treatment
  advances in, 173-212
  combinations of thermal treatment in, 210-212

Heat treatment (continued)
  equilibrium phase diagram and, 174
High-power lasers, welding with, 111–128
High-speed cutting, laser beam in, 49–58
High-speed steel, laser surface hardening and, 60
High temperature alloys, laser weld characteristics for, 123
Hole drilling, with laser beams, 59

Illinois, University of, 278
Induction furnace
  velocity/coil current relationships in, 323
  velocity field in, 327–328
Induction stirring, principle of, 321–329
Instantaneous average cooling rate, dimensionless, 18
Intercritical annealing, 175, 191–197
  of dual-phase steel, 196
Intercritical tempering, 175–191
  austenite distribution in, 187
  austenite precipitation in, 187
Interdendritic flow, 277
  modeling refinements and, 306
  rotation in, 302–305
  in simulated ESR ingot, 287
  slow-creeping, 306
  solute accumulation with, 288
  tortuosity factor in, 307, 311
Interdendritic liquid
  boundary conditions for solving flow of, 286
  density of, 279, 299–301
  flow calculations for, 281
Internal stress superplasticity, 142–149, see also Superplasticity
Ion implantation, in semiconductor crystals, 64

Iron splats, crystalline solidification of, 24–26
Isotherms, shape and location of, 38

Kinetic equation, in plasma dynamics, 335

Laboratory Feasibility Model, in EHD generation of microdroplets, 82–84
Laser(s)
  heat-treating parameters for, 47
  in high-speed cutting, 49–58
  industrial applications of, 112
  operational regimes for, 46–47
Laser-annealed boron implanted silicon, 64–65
Laser beam, in laser machining, 68–69
Laser butt welds, autogenous, 121, see also Laser welding
Laser-controlled surface reactions, 70
Laser cutting, gas jet requirements in, 51
Laser hole drilling, 53, 59
Laser machining, 68–69, see also Material removal process
Laser-melted M2 tool steel, 57–58
Laser performance hardening summary, 48
Laser processing, 45–75, see also Materials processing
  in aerospace industry, 72
  applications of, 71–74
  automatic applications of, 71–72
  beginnings of, 45

Laser processing (continued)
  bulk rapid solidification in, 61-64
  chemical vapor deposition and, 70
  current developments in, 59-72
  laser machining and, 68-69
  in lead welding process, 72
  of M2 high-speed steel, 60
  operational regimes for, 53
  pulse annealing in, 64-65
  in shock hardening, 65-67
  state-of-the-art in, 46-59
  surface melting and, 59-60
  transformation hardening and surface alloying in, 46-49
Laser surface melting
  melt depth and absorbed power in, 54
  schematic for, 54
Laser systems, multikilowatt cw, 112
Laser welding, 49, 111-128
  of aluminum alloys, 122
  deep penetration in, 115, 117
  of ferrous alloys, 117-120
  filler metal addition in, 123-124
  fusion zone purification in, 119, 124-127
  hardness in, 124
  of high-temperature alloys, 123
  impact test results in, 119-120
  industrial applications of, 128
  melting efficiency in, 115
  penetration attainable in, 113-114
  performance and capability in, 114-116
  of titanium alloys, 121-122
  in three-section thickness of HY-130 steel, 126-127
  unique metallurgical characteristics of, 124-128
  weld characteristics and properties in, 117-124
  in X-80 alloy, 50-51
Latent heat of fusion, 348-349

Layerglaze process, 62
Lead-acid storage batteries, laser welding of, 72-73
Liquid-solid interfaces, between uniform and Gaussian heat flux distributions, 40
Local solute redistribution equation, in macrosegregation theory, 280-282
Lorentz force, 321
LSRE, see Local solute redistribution equation

Machining experiments, design of for mathematical modeling, 267
Macrosegregation
  analysis of, 279-283
  dimensionless parameter in, 294-295
  freckle formation in, 294-298
  and local average composition of solid, 283
  local solute redistribution equation in, 280
  minimizing of, 277-313
  mushy zone in, 279-280, 288-290, 293-295, 308-309
  mushy zone energy equation in, 306-307
  permeability and, 307-312
  segregation control in, 298-306
  severity of, 279
  solidification parameters and, 287-294
Magnetohydrodynamics
  liquid melts and, 320-324
  in metals processing operations, 319-341
Maraging steels
  microstructure of, 206-208
  reversion treatment of, 204-209
  strength-toughness combinations in, 204-206, 209

Martensite
  austenite retention in, 186
  cleavage plane in, 185
  dislocated, 184
Martensite packets, decomposition of, 181
Martensitic steel
  intercritical tempering and, 176–177
  reversion of to austenite phase, 176
  structure of, 183
  supersaturated carbon and, 179
Massachusetts Institute of Technology, 278, 339
Material removal process, *see also* Materials processing; Metals processing
  chip removal in, 259
  computer-aided systems in, 257–258
  cost-effectiveness of, 275
  data collection and analysis vs. model types in, 274
  deterministic models in, 261
  economic models for, 267–275
  empirical approach to, 259–265
  empirical models in, 261–267
  historical developments in, 263
  initial machining experiments in, 266
  macro-economic models in, 271–272
  mathematical models for, 257–275
  phenomenological approach to, 259–260
  planning system block diagram for, 273
  probabilistic models of, 262
  statistical models of, 262
  work center in, 269
  working region for end milling in, 273
Material removal system
  feasibility region construction in, 266
  micro-economic model in, 275

Material removal system (continued)
  problems in, 265
  stochastic and dynamic models of, 262
Materials, laser processing of, 45–75, *see also* Laser processing; Laser welding
Materials cycle
  interactions in, 2–3
  materials processing as transition steps in, 8
  processing in, 1–4
  total, 3–4
Materials processing, 1–11, *see also* Laser processing; Powder processing
  micro-economic model in, 272
Materials science and engineering, information flow in, 6
Maxwell's equations, 320, 336
Melt depth
  dimensionless, 33
  fractional, 34
  heat flux and, 33–34
  in laser surface melting, 54
  during melting and crystalline solidification, 32
Metal powder generator
  development of, 79
  tabletop vs. laboratory feasibility model
Metal powders, *see also* Powder processing
  rapidly solidified, 106–107
  secondary electron images of, 83–84
Metals processing, *see also* Laser processing; Materials processing
  arcs used in, 334–341
  electrohydrodynamic techniques in, 79–88
  magnetohydrodynamic and plasma dynamics in, 319–341

Metal substrate, heat flow during solidification against, 22-27
MHD, see Magnetohydrodynamics; see also Electrohydrodynamic metal processing
Michigan, University of, 358
Microdroplets, source for generation of in EHD apparatus, 81
MITAS heat transfer code, 357
Mold design evaluation, solidification simulation in, 365-366
SME, see Materials science and engineering
Mushy zone
   energy equation in, 306-307
   in macrosegregation, 279-283, 288-290, 293-295
   temperature distribution in, 309
   velocity factor in, 308

National Aeronautics and Space Administration, 357
National Science Foundation, 278
Newtonian solidification, dimensionless interface velocity for, 20, see also Solidification
Nickel-base alloy, superplastic forming of, 134-135
Nickel-molybdenum alloy, liquidus/solidus isotherms in production of, 291-292
Non-Newtonian solidification, particle size and, 22
Normalized net solidification time, for liquid metal droplets, 20

Ohm's law, 320-321

Particle size, Bradley's universal plot for determination of, 104

Particulate metallurgy, fundamentals of, 91-107, see also Powder processing
Pendulum grinding, 241
Permeability
   of aluminum alloys, 311
   macrosegregation and, 307-312
   of partially solidified alloys, 313
Plasma, basic description of, 335
Plasma dynamics
   and arcs used in metals processing, 334-341
   in electric arc welding, 341
   kinetic equation in, 335
   in metals processing operation, 319-341
Plasma flow field, for free-burning arc, 337
Poisson's equation, 335
Polycrystalline materials, tensile deformation of, 133
Potential process developments, examples of, 11
Powder metallurgy, see also Metal powders; Powder processing
   high-performance, 91-92
   particle-size distribution in, 95
   powder production and processing in, 91-92
   rapidly solidified powders in, 106-107
Powder processing
   atomization in, 92-106
   centrifugal atomization in, 97-99
   cooling rate in, 91
   dendrite arm spacing vs. cooling rate in, 106-107
   materials amenable to, 91
   rapidly solidified powders in, 106-107

Powder processing (continued)
  roller atomization in, 99, 105–106
  rotating electrode process in, 97–98
  steam atomization in, 96–97
  ultrasonic atomization in, 99, 105
  vacuum or soluble gas atomization in, 99
  water atomization in, 95–97, 103–105
Prandtl number, 14
Process developments, examples of, 11
Processing, *see also* Laser processing; Materials processing; Metals processing; Powder processing
  classes of, 7–11
  materials cycle in, 1–4
  materials science and engineering in, 4–6
Pseudo-steady state technique, 353–355
Pulse-annealed ion implanted silicon, 62–63
Pulse hardening, laser beam in, 64–65

Rapid laser surface-melted alloys, 55
Rapidly solidified metal powders, 106–107, *see also* Powder processing
Rapid solidification processing
  critical cooling rate during, 25
  defined, 13
  heat flow limitations in, 13–42
  structure in, 107
Rapid surface melting, laser beam and, 61
Reheating of steel, responses to, 174, *see also* Steel(s)

Rewelding, laser shock hardening in, 67
Reynolds number, 14
Roller atomization, mechanism of, 105–106
Rotating electrode process, in centrifugal atomization, 97
Roughness detection, new method in, 248–249
RSP, *see* Rapid solidification processing
Ruby laser, 111, *see also* Laser(s)

Sand ingot castings, 357–360
Secondary electron images
  of metal powders, 83–84
  of splat-cooled Al-Cu specimens, 86–88
Segregation, inverse, 280
Segregation control, in macrosegregation, 298–306
Self-consistent field, in plasma dynamics, 335–336
Semiconductor crystals, laser radiation in, 64
Sheet of liquid, disintegration of, 102
Ship steel, laser tee weld in, 118
Shock hardening, laser beams in, 65–67
Silicon carbide, as grinding material, 215
Simulation, thermal properties and, 346
SINDA heat transfer code, 348, 357
Slip creep, 157, 163
Solidification
  of aluminum–silicon alloys, 359
  computer simulation in, 345–367
  fluid flow simulation in, 355–356
  heat flow during, 22–27

INDEX

Solidification (continued)
  and heat of fusion, 347-350
  heat transfer and, 347-355
Solidification of droplets,
    heat flow during, 14
Solidification parameters,
    macrosegregation and,
    287-294
Solidification problem, modeling
    of, 356-358
Solididication simulation
  for aluminum cylinder block
    casting, 361
  applications of, 358-366
  for centrifugal casting, 362
  computer in, 345-367
  in feeding distance calcula-
    tions, 365
  fluid flow in, 355-356
  mold design evaluation in,
    365-366
  for sand ingot castings,
    357-360
  for special casting processes,
    362-363
  in welding, 363-365
Specific heat method, heat of
    fusion and, 348
Speed stroke grinding, 241-244,
    see also Grinding
Splat cooling, with EHD
    apparatus, 85-87
Spot welding, see also Laser
    welding; Welding
  numerical techniques in,
    364-365
  with pulsed laser systems,
    111-112
Stainless steel, see also
    Steel(s)
  corrosion susceptibility of,
    56
  laser weld penetration in, 114
  sensitization diagram for, 55
  source of for EHD generation
    of microdroplets, 81
Steam atomization, in powder
    processing, 96-97
Steel(s)
  austenite reversion of, 176

Steel(s) (continued)
  combinations of thermal
    treatments for, 210-212
  dual-phase, see Dual-phase
    steels
  ductile-brittle temperature
    transition in, 177
  heat treatment of, 173-212
  intercritical annealing of,
    175
  intercritical tempering of,
    175-191
  maraging, 204-206
  normal tempering of, 175
Steel conditioning, machines
    and grinding wheels
    for, 216
Steel toughness, retained
    austenite and, 181
Strain site sensitivity
  at high temperatures, 139
  superplasticity and, 165
  thermal cycling and, 144-145
  for $\alpha$-uranium, 143
Substrate, surface temperature
    of, 33
Superplastic behavior,
    temperature for, 154
Superplastic flow
  lattice diffusion-controlled,
    158
  optimizing rate of, 157-161
  stress dependence on, 149
Superplastic forming
  of nickel-base alloy, 134-135
  of titanium-aluminum alloy,
    136
Superplasticity, 133-167
  applied mechanics aspects
    of, 138-141
  creep flow and, 143-147
  defined, 133
  fine structure, 149-161
  internal stress, 142-149
  materials science aspects of,
    142-165
Superplastic materials, S-
    shaped behavior in,
    153

Superplastic powders, densification of, 161, 164
Superplastic properties, application of, 134-138
Superplastic solid state bonding, 161-165
Superplastic steel plates, solid-state diffusion bonding of, 166
Surface hardening, lasers in, 46-49
Surface layer, heat flow during melting and solidification of, 28-42
Surface melting, laser beams in, 59-60
Surface roughness, in-process detection of, 248

Tabletop rapid quenching metal powder generator, 79
Taylor's expansion theorem, 351
Temperature distributions, dimensionless, 16
Thermal cycling, strain rate sensitivity and, 144
Thermal properties, electrical equivalents for, 350
Thin-gage materials, welding of, 113
Titanium alloys, laser weld characteristics in, 121-122
Titanium-aluminum alloy, superplastic forming of, 136
Tool life, experimental designs in, 262-265
Tool steel M2, laser-melted, 57
Tortuosity factor, in interdendritic flow, 307
Transformation hardened grey cast iron, cross section of, 48
Transformation hardening, lasers and, 46-49
Transformation strain, in fine structure superplasticity, 150

Transformation strengthening, of austenite, 199-204
Transient heat transfer code, 358
TRUMP heat transfer code, 357
Two-B cycling treatment, for cryogenic steels, 210-212

Ultrasonic atomization, mechanism of, 105
α-Uranium
  creep behavior of, 143
  thermal cycling of, 144

Vacuum induction melter, 330
Velocity field, in induction furnace, 324-328

Water atomization
  mechanism of, 103-105
  operating conditions in, 96
  in powder metallurgy, 94-97
  "scrape" mechanism in, 104
Welding
  deep penetration, 113
  electron beam, 364
  electroslag, 329-334
  laser, 49, 111-128
  multikilowatt systems in, 112
  solidification simulation in, 363-365
Weld simulation, finite difference and finite element techniques in, 364

X-80 alloy, laser welds in, 50-51

YAG (yttrium aluminum garnet), 112
YALO (yttrium aluminate), 112

Zirconia, co-fusion with alumina, 215, *see also* Alumina-zirconia alloys